Women, Gender, and Technology

WOMEN, GENDER, AND TECHNOLOGY

Series Editors
Mary Frank Fox, Georgia Institute of Techology
Deborah G. Johnson, University of Virginia
Sue V. Rosser, Georgia Institute of Technology

Women, Gender, and Technology

Edited by

MARY FRANK FOX,
DEBORAH G. JOHNSON, AND
SUE V. ROSSER

UNIVERSITY OF ILLINOIS PRESS
Urbana and Chicago

Library of Congress Cataloging-in-Publication Data
Women, gender, and technology / edited by Mary Frank Fox,
Deborah G. Johnson, and Sue V. Rosser.
p.   cm. — (Women, gender, and technology)
ISBN-13: 978-0-252-03095-6 (cloth : alk. paper)
ISBN-10: 0-252-03095-8 (cloth : alk. paper)
ISBN-13: 978-0-252-07336-6 (pbk. : alk. paper)
ISBN-10: 0-252-07336-3 (pbk. : alk. paper)
Includes bibliographical references and index.
1. Women—Effect of technological innovations on—United States.
2. Computers and women—United States.
3. Information technology—Social aspects—United States.
I. Fox, Mary Frank.
II. Johnson, Deborah G., 1945–
III. Rosser, Sue Vilhauer.
HQ1178.W68   2006
303.48'34082—dc22   2006018048

Contents

Preface

During the year 1999–2000, the three of us found ourselves together in a reading group on gender and technology, sponsored by the Women, Science and Technology Center (WST) at the Georgia Institute of Technology. The reading group consisted of faculty and staff from a wide variety of places and fields across the campus—engineering, economics, counseling psychology, literature, philosophy, and sociology. The group discussion was exciting, but we all were surprised to find a paucity of materials on gender and technology.

Despite the increasingly rapid development of women's studies and technology studies, the fields have tended to progress in parallel, with some crossover between them—though mostly from women's studies to technology studies, and less the other way. To be sure, the literatures on science and gender and feminist science studies are relevant and relatively abundant, but they do not quite address what is central to the relationship of gender and technology. A good deal of literature also exists on the participation of women in both science and engineering, and assesses barriers to the entry of women into these fields. In addition, science fiction is thoughtful on the topic. But scant women's studies scholarship has focused on women and technology, feminist perspectives on technology, and/or the gendering of technology and its impact upon gender relations in society. Technology studies have produced a handful of books focused on gender and technology, although these have been by non-American authors. Among other things, what are missing are multiple approaches to and perspectives on women, gender, and technology from different levels of analysis: culture and societies, institutions and organizations, and individuals.

Hence, we launched the idea of a book series that will encourage, facilitate, and bring to an interdisciplinary audience such a range of theory, research, and applications on women, gender, and technology. This book is the first in that series; several chapters are contributions from women who participated in the reading group.

We anticipate that this volume will be the first in a long and fruitful series. We hope the topic of women, gender, and technology flourishes into the next decade—and that this series is part of its flowering.

Women, Gender, and Technology

Introduction

DEBORAH G. JOHNSON

In popular media as well as scholarly literature, contemporary American society is often characterized as "technological" or "techno-scientific." We are told that we live in an era of unprecedented technological change, and that technological development and innovation will only escalate in the twenty-first century. Any good feminist, it would seem, would greet these platitudes with questions about what it all means for women. Does more and better technology necessarily mean better for women? How does technology (or how do particular technologies) affect gender relations? Will the technologies of the twenty-first century improve the lives of women?

Until a decade ago, questions about the gender–technology relationship rarely received attention. Today, increasing interest exists. To be sure, feminist scholars began to examine the science–gender relationship several decades ago. Evelyn Fox Keller's book on the life of Barbara McClintock was groundbreaking when it appeared in 1984, and Sandra Harding's *The Science Question in Feminism* stirred further discussion and debate when it appeared two years later. The literature on science and gender is important for understanding gender and technology, but it does not and cannot address the distinctive character of technology.

Studies of gender and science have focused on science both as a system of knowledge and as a social institution. Theorists have noted and tried to understand and explain why there are so few women in science. In turn, they have questioned whether and how the lack of women in science affects the knowledge produced. In the case of technology, the parallel phenomenon is engineering, and the parallel questions are about the number of women in engineering, and whether and how this affects both the production of engineering knowledge and the products of engineering. However, technology is much more than engineering. While technology often originates with engineers, many other actors and institutions are involved in determining which technologies succeed, how technologies are used, and what cultural meaning is associated with them. Most important, technology is associated with material objects (artifacts), and material objects influence human activities through their materiality. We encounter technology as we move physically and socially through our lives. Sometimes technology facilitates what we do; other times it constrains what we do. Theorists of gender and technology must come to grips with the materiality and the facilitating and constraining features of technology in addition to the character of engineering.

Nevertheless, technology is not just artifacts. Social studies of technology emphasize that technology is better conceived as a system comprised of artifacts, social practices, and systems of knowledge. On these accounts, artifacts are inseparable from the social meanings and practices associated with them. When technology is understood in this way, it is easier to see how gender might be intertwined with technology. That is, gender relations and ideas about gender may be intertwined in the social practices and meanings associated with particular artifacts. Understanding gender and technology can be a matter of understanding how gender comes to be embedded and carried in the design and meaning of technological artifacts as well as in the uses of such artifacts. This understanding can be informed by, but is still distinct from, understanding gender and science.

Moreover, science and engineering are different communities with distinctive values, practices, histories, and cultures (Layton, 1971). Thus, while scholarship on science and gender can make a contribution to understanding gender and technology, theorists of gender and technology must come to grips with the values, practices, and culture of engineering, with its emphasis not just on knowing but on doing. Much more attention needs to be given to this topic.

The current interest in gender and technology arises from several directions. We have included "women" in the title of this volume because of its connection to women studies and feminist scholarship, focusing on women as subject and bringing women to the forefront of study. From the perspective of women studies, technology and engineering represent one of the last bastions of male domination and appear somewhat impervious to gender change. The relationship between gender and technology is indubitably part of the story that women studies/feminist literature seeks to tell, the story explaining the barriers to equality of both sexes and the cultural meaning of female, feminine, and woman.

A survey of the most prominent work on gender and technology suggests that interest in gender and technology is also coming out of the field of science and technology studies (STS). (See Wajcman, 1991, and Wajcman, 2000, for discussion of the state of the literature.) STS scholars uncover and document the ways in which science and technology are socially shaped, and gender comes into focus as an aspect of the social that is at once shaped by and shapes science and technology. Indeed, one of the early and now classic works in the field was an analysis of the evolution of the design of the bicycle. Bijker (1995) argues that the development of the bicycle did not follow a linear path, and the design we know today endured ("stabilized," in his terms) because multiple interest groups became attached to it, albeit for different reasons and with diverse meanings associated with the design. In discussing struggles over the meaning of various models, Bijker notes that one of the models, the Ordinary, was considered unsafe by some groups, and the dangerous aspects of the model served the interests of another group, young men who would frequent public parks. The "young and often upper-class men could display their athletic skills and daring by showing off in the London parks," Bijker explains. "To impress the riders' lady friends, the risky nature of the Ordinary was essential. Thus the meanings attributed to the machine by the group of Ordinary users made it a Macho Bicycle" (pp. 74–75). In this way, Bijker implicitly argues that gender influences the meanings associated with a design and ultimately influences which designs succeed and which disappear.

The early literature on gender and technology suggests that gender is coded into many technologies in complex ways. It suggests that gender and technology are profoundly intertwined—in the deep structure of our cultural conceptions. Consequently, the study of gender and technology requires a massive interdisciplinary effort. Among STS scholars and others, some consensus has developed around a co-creation thesis—gender and technology co-create one another. As suggested a moment ago, technology is socially shaped; gender patterns in society can therefore be reproduced in constituting technology. At the same time, technology shapes society: if gender has been coded into a technology, that technology may reinforce gender patterns.

The co-creation thesis is a powerful idea insofar as it is, at once, both simple and enormously complex. This is precisely what is needed in a new and widely interdisciplinary field. The thesis has content and yet does not highly constrain further inquiry. It provides the broad contours of an account and calls for the details to be filled in by further (multiple disciplinary and interdisciplinary) analyses. The co-creation process must be described and explained. Cases in which it does not occur, as well as those in which it occurs, need to be examined. The thesis goes well beyond naming the area of study, though even naming is an important step in the development of new knowledge.

"Cocreation" is shorthand for two ideas. The "co" signals valence in both directions: gender affects technology *and* technology affects gender. "Creation," the more mysterious term, signals that gender and technology shape and affect one another, and that one can be expressed in and through the other. Authors use a variety of verbs to describe the creation process; gender and technology may *create, define, redefine, reinforce, shape, affect, influence, perform,* or *express* one another. The list is endless, as every imaginable perspective might be used to explore the relationship between gender and technology. In a sense, the agenda set out by the co-creation thesis is to specify and explain these and other verbs. We want to understand how and why gender and technology are connected; how they do and do not affect one another; how to describe and explain their workings on each other.

The cocreation thesis points in a number of useful directions, suggesting gender at work in many aspects of technology. We find gender in the processes by which technologies are developed (e.g., in the institutions, relationships, and ideas that produce and distribute technologies). We find it in the products themselves (e.g., encoded in design features and in the meanings attached to technological objects). And we find gender in the use of technologies (i.e., in who has access and who adopts technologies). Building on the co-creation model, Juliet Webster (1996) used these three themes in her analysis of the information technology industry, and in chapter 1 of this volume Rosser uses a parallel distinction, focusing on women in the technology workforce, women and technology design, and women as users of technology. This threefold distinction moves the co-creation model forward by telling us where to look for the action of gender and technology.

The "co" tells us to expect technology lurking in gender relations and vice versa. This may seem too obvious to be worthy of mention since the association between masculinity and technology is so strong in Western culture. However, exploring what seems obvious points to one of the obstacles of working in this area. Gender and technology are at once both intimately connected and fluid terms. Consider how we typically think about the gendered division of labor in the household. Men, so we stereotypically think, have responsibility for the household tasks involving technology—taking care of the automobiles and home repairs. Women handle the apparently nontechnological tasks such as cooking and childcare. At first blush it seems that we can say, "see, technology is 'there' in the gender relations." Of course, to make such a claim requires a bit of contortion about what counts as a technology. Aren't pots and pans, brooms, and childrearing know-how technologies? A task cannot be technological simply because it involves the use of tools: mopping the floor, cooking, and changing diapers all involve the use of tools. So, does the association between technology and masculinity lead to the division of household tasks along gender lines, or do gender relations lead to the classification of those tasks as technological and nontechnological? (See Croissant, 2000, for a discussion of what does and does not count as technological.) Perhaps, the status of the group doing a task determines whether the task is defined as technological. If this were true, then we would expect to find something parallel happening with tasks performed by racial/ethnic groups as well. In any case, the fluidity and inextricability of gender and technology make for enormous complexity; the co-creation thesis begins to deconstruct this complexity.

While the co-creation model and its elaboration frame our understanding of gender and technology, the task remains daunting. We have identified *the beast,* named it, and seen enough to know that we must constantly keep it in our sights. The beast seems impenetrable even though (and perhaps because) we are part of it (and it is part of us). We want to understand the workings of cultural conceptions—categories, ideas, meanings, notions—about gender and technology, and yet we are forced to *use* those very concepts to do the analysis. Not only is the beast complicated, but we must view it both through its own eyes as well as from some other place. We get all tied up in knots. Add to this that we aim not just to understand the beast; we also want to change it.

All of the works in this collection aim, in one way or another, to understand the co-creation of gender and technology. Each chapter explores how gender and technology work and are at work in a particular domain or expression such as in film narratives, reproductive technologies, the digital divide, and the profession of engineering. The authors of chapters represent more than a half dozen different disciplines, reflecting the need for multiple and interdisciplinary approaches to understanding.

We begin with Sue Rosser's reflections on how different feminist theories can be used to illuminate different aspects of gender and technology. "Feminism," like "gender" and "technology," is a fluid and somewhat contentious term. Rosser's method reflects the diversity and complexity of feminist approaches; she uses these diverse perspectives to illuminate issues in gender and technology. These theories do not point to a single question or problem but to a rich set of insights and questions, some of which are addressed in later chapters.

Rosser's section on African American/Womanist and Racial/Ethnic Feminism introduces a theme of race and technology quietly woven in this volume. Race and technology seem to be as inextricably intertwined as gender and technology, although the association is less often recognized, let alone understood. Barbara Katz Rothman (chapter 6) suggests this intertwining in her discussion of genetic technology, and Cheryl Leggon (chapter 5) addresses it head-on in her analysis of the digital divide. The theme points to a direly underdeveloped area of study and reminds us of an important lesson about technology: it affects different groups of people (as well as different individuals) differently.

Engineering is a key focal point for studies of gender and technology because engineers, acting individually and collectively in a variety of institutional contexts, develop and deploy technology. Mary Frank Fox (chapter 2) looks at the educational and career paths of men and women with PhDs in engineering as she documents the profile of gender, status, and engineering, the challenges and obstacles for women, and prospects for the future. Mara Wasburn and Susan Miller (chapter 3) provide a case study of a program aimed at retaining women enrolled in a school of technology at a major engineering university.

Academic institutions train engineers and technical specialists who design, produce, maintain, and distribute the technologies that, in turn, shape and are shaped by gender relations. Gender differentiation permeates academe. Both the chapters by Fox and Wasburn and Miller recognize that technology and engineering are culturally correlated with masculinity. As Fox puts it, "Images, symbols, and systems of belief have continued to link engineering with men and masculinity and separate it from women and femininity" (p. 54). Both chapters confront male domination in technical fields. Fox meticulously describes the phenomenon and seeks to identify social and organizational factors that keep the numbers of women in engineering low. Wasburn and Miller represent a more activist thread. Acknowledging that social and organizational factors affect the success of women in technical education, they search for a program—a social organizational mechanism—to counteract the built-in obstacles and to help women be successful in undergraduate programs.

The Wasburn and Miller account is followed by Judy Wajcman's examination in chapter 4 of the changes in work brought about by information and communication technologies (ICTs) and their meaning for women. Remember that the co-creation thesis does *not* claim that technology always reproduces or reinforces prevailing gender relations. Just as gender relations may affect the design of technology, technology might affect gender relations—for better or worse. Thus, the co-creation thesis is compatible with the possibility that ICTs will democratize and equalize gender as well as other social relationships. Wajcman takes this possibility seriously in her examination of how women are being affected as ICTs alter workplaces.

Wajcman frames the question in terms of what the shift to a service economy means for women's position and experiences in the labor market and, in particular, its effects on women's careers and gender power relations in management. She examines this on an international scale, drawing mixed conclusions. The changes being brought about in the workplace by ICTs "can be constitutive of new gender power dynamics but they can also be derivative of or reproduce preexisting conditions of gender inequality at work" (p. 94).

Continuing with a focus on information technology and what it means for women, Leggon, in chapter 5, considers how African American and Hispanic women will be broadly affected by information technology. Leggon provides a provocative analysis not just of gender and technology but also of the intertwining of race and ethnicity with gender *and* technology. Leggon examines data that reveal patterns of behavior of black and Hispanic women as producers as well as users of information technology. The data suggest a picture much more complex and changing than the simplicity implied by "the digital divide." We find a story of the intertwining of race, ethnicity, gender, and technology involving, among other things, attitudes toward African American and Hispanic women that covertly shape the way technologies are developed and used, and whom the technologies serve. Leggon concludes with a discussion of the implications of her analysis for policy debates that have traditionally excluded the interests of African American or Hispanic women.

Following Leggon's analysis are two chapters pursuing the theme of gender at work in the development and use of genetic science and reproductive technologies. Genetic science has led to the development of many technologies, technologies that are generally thought to address problems that afflict women. Chapters 6 and 7 each present a more complicated picture of the reality.

In chapter 6, Rothman provides a deep analysis of the cultural meaning of genetic science. This calls for an informal, interpretive account of cultural conceptions. Rothman's analysis is a version of the co-creation thesis: she argues that genetic science and the technologies developed from it "grow out of and maintain traditional gendered divisions." In other words, gender divisions shape the science and technology and then reinforce the divisions. The most fundamental ideas of genetic science—its ideology, its logic—carry gender differentiation, she argues. Rothman sees gender (in this case, patriarchy) at work in the way we define the real mother as new technologies allow eggs to be moved from woman to woman. The old patriarchal idea that seed (given by the father) determined who the father was (and who the child was) is applied to women. Now that eggs can be moved around we use a parallel model presuming that the "real" mother is the one who provided the egg. In this way, patriarchy continues to shape ideas about who the "real" mother is and to whom the baby belongs.

Rothman argues that while the *rhetoric* around new genetic technologies suggests empowerment for women, the *results* of these technologies have been far from empowering to women. Although genetic testing has made it easier for women to decide to get an abortion, it has diverted us from changing social conditions that might tip the balance in favor of keeping babies.

Consistent with the co-creation thesis, Rothman suggests that social attitudes toward nontraditional women (e.g., lesbian couples) played a role in changing the direction of technological development. Research and rhetoric changed when it was recognized that the new technologies facilitated nontraditional women in having babies. In pointing this out, Rothman gives further support to Leggon's idea that attitudes toward new technologies and directions of development are affected by the group whom the technology seeks to help. Finally, Rothman argues that the portrayal of breast cancer as a genetic disease is not only misleading but does not serve women well; framing breast cancer as a genetic disease serves the interests of genetics laboratories and pharmaceutical companies, scientists, and researchers, as well as corporations that pollute the environment.

In chapter 7, cultural anthropologist Linda Layne presents the results of her ethnographic studies of the effects of new reproductive and information technologies on women's experience of miscarriage, stillbirth, and early infant death. Layne documents changes in the incidence and causes of pregnancy loss as well as changes in the expectations of women and men with regard to pregnancy. New reproductive and information technologies have changed prenatal bonding and allowed pregnancy loss to be identified earlier in the course of pregnancy. While these technologies might be seen as unqualifiedly beneficial from medical and emotional perspectives, Layne's data and analysis show the effects of the technology to be much more complicated and profound in their influence on the way women now experience pregnancy.

Layne thinks of many of the effects as fitting the pattern that Edward Tenner (1996) describes as "technology biting back." She also uses Langdon Winner's notion of technological somnambulism (1986). Winner is struck by how people "so willingly sleepwalk through the process of reconstituting the conditions of human existence" when new technologies are introduced. Layne concludes by pointing to the need for more attention to the issue:

> [D]espite twenty years of social scientific research on the impact of new reproductive technologies, and despite the women's health movement's emphasis on empowering women through increasing their control over their bodies, a host of new technologies has significantly changed the experience of pregnancy loss in unintended *and* unexamined ways. Even in the area of reproductive medicine, where one would expect to find very low degrees of technological somnambulism, given feminists' keen attention to these matters, in the case of pregnancy loss . . . we seem to be in a deep, deep sleep.

Given that cultural conceptions affect access to, development of, and use of technology—as saliently illustrated in both Leggon's and Rothman's analyses—we should expect to find the connections between gender and technology expressed in literary works. In chapter 8, Carol Colatrella examines these connections in two film narratives in which the authors creatively play with the gender and technology connection. Both films, *Eve of Destruction* and *Making Mr. Right,* explore what happens when women are involved in developing robots, a role typically associated with men. In effect, the film narratives explore an idea frequently asked by feminists interested in science and technology: would science and its technological products be radically different if many more women were doing science and engineering?

Colatrella places these two narratives in the backdrop of a review of STS theory and literary works expressing a variety of (sometimes even contradictory) attitudes and ideas about the influence of women on technology and vice versa. One of the intriguing themes that Colatrella points to is a connection between the involvement of women with technology and more ethical uses of it.

These narratives illustrate how we sometimes must use the very ideas we seek to change, for the authors Colatrella discusses play on and with gender stereotypes in order to explore alternatives. The messages uncovered in Colatrella's analysis are rich, complex, and intricate. In the *Eve of Destruction,* for example, she finds a combination and reworking of both the Frankenstein theme and the Cinderella story together with the traditional robot theme of humans attempting to recreate something in their own image. The narrative of *Making Mr. Right* is also complex but more comical in the messages it gives about gender and technology. Here the central figure, a woman named Frankie, teaches social skills and graces to a male robot to make him more endearing to the public.

Technology is not traditionally associated with religion. Thus, chapter 9 by James Fenimore would be intriguing if it only gave a glimpse of how new multimedia technologies are entering the domain of religion, elevating the profane to the sacred. However, the account is even more interesting (and appropriate for this volume) because along with the new technologies come gender associations. Fenimore's account shies away from claiming that these new technologies "carry" gender, but they are, at least, new platforms for the reproduction of gender ideas. As churches use multimedia technologies, their use and the rhetoric and images around their use are highly gendered. In this way, Fenimore's chapter supports the co-creation thesis. Gender shapes how the new multimedia technologies are used; gender is used to send religious messages about the technology. Adoption of the new technologies reinforces gender politics. Since evangelical politics are highly gendered, perhaps we should not be surprised to find the adoption of new technologies to follow suit.

The nine chapters in this volume provide a picture of the rich and complex issues that come into play in trying to understand the gender–technology relationship. We see gender influencing and being influenced by technology in a wide range of contexts, including engineering and engineering education, the workplace, medicine, literature, and religion. These chapters take us a step closer to understanding a territory that will continue to compel more attention.

References

Bijker, W. E. (1995). *Of bicycles, bakelites, and bulbs: Towards a theory of sociotechnical change.* Cambridge, Mass.: MIT Press.

Croissant, J. (2000). Engendering technology: Culture, gender, and work. In Shirley Gorenstein (Ed.), *Knowledge and society: Vol. 4. Research in science and technology studies: Gender and work* (pp. 189–207). Stamford, Conn.: JAI Press.

Harding, S. (1986). *The science question in feminism.* Ithaca, N.Y.: Cornell University Press.

Keller, E. F. (1984). *A feeling for the organism: The life and work of Barbara McClintock.* New York: Henry Holt & Company, Inc.

Layton, E. (1971). Mirror-image twins: The communities of science and technology in 19th-century America. *Technology and Culture 12,* 562–580.

Tenner, E. (1996). *Why things bite back: Technology and the revenge of unintended consequences.* New York: Alfred A. Knopf.

Wajcman, J. (1991). *Feminism confronts technology.* University Park: Pennsylvania State University Press.

Wajcman, J. (2000, June). Reflections on gender and technology studies: In what state is the art? *Social Studies of Science 30/3,* 447–464.

Webster, J. (1996). *Shaping women's work: Gender, employment and information technology.* London: Longman.

Winner, L. (1986). *The whale and the reactor.* Chicago: University of Chicago Press.

1

Using the Lenses of Feminist Theories to Focus on Women and Technology

SUE V. ROSSER

The case studies and increasing numbers of well-documented interactions of women with specific technologies appear to be ready for examination using a broader spectrum of feminist lenses to reveal insights such as those found for women and science (Rosser, 1993), women's health (Rosser, 1994), and women's relationship to bio- and reproductive technologies (Rosser, 1998, 2000). As Juliet Webster (1995) notes, feminist analyses of technology have typically fallen under the three categories of liberal feminism, ecofeminism, and, most particularly, socialist feminism. Webster and others (Cockburn, 1981, 1983, 1985; Hacker, 1981, 1989; Wacjman, 1991) may have favored socialist feminist approaches because women's relationships with technology typically have been subordinated and excluded from study, including, to some extent, from studies of the social shaping of technology. Scholars exploring gender and technology interactions have tended to concentrate more on gender and technology in the workforce and somewhat less on women as users of technology. Limited studies have explored how technological designs, especially for information technologies (IT) and household appliances, might differ depending on the gender of the designer and user. Although some funded, co-curricular, and pedagogical projects have explored techniques to attract women students and retain them in engineering curricula, none has significantly changed curricular content or affected, as one woman engineer stated, "the fundamentals." The framework provided by the perspective of each feminist theory lends itself to an examination of three areas: women in the technology workforce, women as users of technology, and women and technology design. Examining technology and gender using the fuller range of feminist theories may indeed uncover subtle, rich insights into the dynamics of the co-evolution of gender and technology.

Liberal Feminism

A general definition of liberal feminism is the belief that women are suppressed in contemporary society because they suffer unjust discrimination (Jaggar, 1983). Liberal feminists seek no special privileges for women and simply demand that everyone receive equal consideration without discrimination on the basis of sex.

WORKFORCE

When it comes to technology jobs, most engineers and others involved with technology take a liberal feminist stance and assume that the focus should be on employment, access, and discrimination issues. For example, nationally organized programs such as Women's Engineering Program Advocates Network (WEPAN) and Society of Women in Engineering (SWE), as well as women in engineering programs developed by individual institutions, demonstrate a liberal feminist focus in their attempts to remove documented overt and covert barriers that prevent women from entering engineering education and remaining as practicing engineers. Social scientists studying the gender distribution of the technology workforce point out that historically and presently, the technology workforce represents a vertically and horizontally gender-stratified labor market, with women concentrated in the lowest-paid positions, closest to the most tedious, hands-on making of the products and furthest from the creative design of technology.

In this framework, Cynthia Cockburn's (1983) study of compositors is insightful at showing the way in which gender works to keep women out of technology jobs. Male typesetters tried to retain their high pay by demanding the sole rights to use computer typesetting equipment. In these demands they excluded women and defined them as unskilled. The technological innovation that led to typesetting being equivalent to electronic keyboarding (QWERTY) opened the door for management to replace male Linotype operators with cheaper female typists. Today most women working in the IT industry engage in the tedious, eye-straining work of electronic assembly. Men predominate in the decision-making, creative, and design sectors as venture capitalists, computer scientists, and engineers producing startups, new software, and hardware design.

DESIGN

Liberal feminists would seek to remove barriers that prevent equal access for women to technology jobs not only to ensure economic equality but also to provide access to higher paying jobs for women. Unequal access has implications that go well beyond the composition of the workforce. Two decades ago, colleagues in biology (Birke, 1986; Bleier, 1984, 1986; Fausto-Sterling, 1992; Hubbard, 1990; Keller, 1983, 1985; Rosser, 1988; Spanier, 1982) revealed that a predominance of male scientists had tended to introduce a source of bias by excluding females as experimental subjects, focusing on problems of primary interest to males, employing faulty experimental designs, and interpreting data based in language or ideas constricted by patriarchal parameters. This exclusion led to bias that had particularly problematic consequences in areas such as health, where the bias resulted in underdiagnosis, inappropriate treatment, and higher death rates for cardiovascular and other diseases in women (Healy, 1991; Rosser, 1994).

Male dominance in engineering and the creative decision-making sectors of the IT workforce may result in similar bias, particularly design and user bias. Shirley Malcom (personal communication, March 9, 1998) suggests that the air bag fiasco suffered by the U.S. auto industry serves as an excellent example of gender bias reflected in design; this failure would have been much less likely had a woman engineer been on the design team. Since women, on average, tend to be smaller than men, a woman designer might have recognized that a bag which implicitly used the male body as a norm would be flawed when applied to smaller individuals, killing rather than protecting children and small women.

USE

Having large numbers of male engineers and creators of technologies often results in technologies that are useful from a male perspective (i.e., these technologies fail to address important issues for women users). In addition to the military origins for the development and funding of much technology (Barnaby, 1981; Norman, 1979), which makes its civilian application less useful for women's lives (Cockburn, 1983), men designing technology for the home frequently focus on issues less important to women users. For example, Anne-Jorunn Berg's (1999) analysis of "smart houses" reveals that such houses do not include new technologies; instead they focus on "integration, centralized control and regulation of all functions in the home" (p. 306). "Housework is no part of what this house will 'do' for you" (p. 307).

Knowledge of housework appears to be overlooked by the designers of smart houses. As Ruth Schwartz Cowan's (1976, 1981) work suggests, the improved household technologies developed in the first half of the twentieth century actually increased the amount of time housewives spent on housework and reduced their role from general manager of servants, maiden aunts, grandmothers, children, and others, to an individual who worked alone doing manual labor aided by household appliances.

Many studies have explored the overt and covert links between the military and design and use of technology. For example, Janet Abbate (1999) studied the origins of the Internet in the U.S. Defense Department's Advanced Research Projects Agency Network (ARPANET). The unique improvement of the Internet was that it was a network, overcoming the vulnerability to nuclear attack of the previous star configuration computer network. Sally Hacker (1981), Cockburn (1983) and others (Enloe, 1983; Fallows, 1981; Weber, 1997) have studied and written extensively on the masculinity of engineering and technology and its military origins. In relationship to liberal feminism, I wish to emphasize how the military origins and funding of much development of technology, coupled with the predominance of men designing the technology, influences its use by gender. Although technologies designed for military uses sometimes are used in civilian life, these tools and systems tend to be more useful in the male sphere.

Although liberal feminism suggests that true equity of women in the technology workforce would correct bias in design and better serve women's interests, liberal feminism, by definition, does not address the potential of gender to affect "fundamentals" (i.e., do women engineers define, approach, or discover different fundamentals such as string theory?). Liberal feminism accepts positivism as the theory of knowledge and assumes that human beings are highly individualistic and obtain knowledge in a rational manner that may be separated from their social conditions, including conditions of race, class, and gender. Positivism implies that "all knowledge is constructed by inference from immediate sensory experiences" (Jaggar, 1983, pp. 355–356). These assumptions lead to the belief in the possibilities of obtaining knowledge that is both objective and value-free, concepts that form the cornerstones of the scientific method. Since liberal feminism re-affirms, rather than challenges, positivism, it suggests that "fundamentals" would always remain the same. Now that they have become aware of potential bias in design or user-friendliness, both male and female engineers and technology creators can correct for such biases that previously resulted from failure to include women and their needs and interests.

Socialist Feminism

In contrast to liberal feminism, socialist feminism rejects individualism and positivism. Marxist critiques form the historical precursors and foundations for socialist feminist critiques and define all knowledge, including science, as socially constructed and emerging from practical human involvement in production. Since knowledge is a productive activity of human beings, it cannot be objective and value free because the basic categories of knowledge are shaped by human purposes and values. In the early twenty-first-century United States, capitalism, the prevailing mode of production, determines technology and favors the interests of the dominant class. This Marxist/socialist theory undergirds the work of numerous scholars of technology who have used this framework for their studies, producing a large body of research commonly known as "the social shaping of technology." Scholars suggest not only that technology is a social product, as implied directly by the phrase "the social shaping of technology," but they extend the phrase's meaning to suggest that technology also comprises human activities and "know-how" (MacKenzie & Wajcman, 1985).

As Webster (1995, p. 4) and other feminist scholars (Wajcman, 1991) rightly point out, technology and the social shaping of technology have often been conceptualized in terms of men, excluding women at all levels. Socialist feminist critiques include women, and they place gender on equal footing with class in shaping technology. In this dual systems approach (Hartmann, 1981; Eisenstein, 1979) capitalism and patriarchy function as mutually reinforcing parts of a system where the sexual division of labor stands with wage labor as a central feature of capitalism, and where gender differences in wages, along with failing to count contributions of women to reproduction and child rearing as "productivity" in a capitalist economy, reinforce patriarchy and power differentials in the home.

WORKFORCE

As Webster suggests, feminist approaches to the social shaping of technology, also called feminist constructivist approaches (Gill & Grint, 1995), and which I call socialist feminist approaches here, have been useful. They are probably the most widely developed and used of the feminist theoretical approaches, framing studies of women and technology at all different levels. The dual system of socialist feminism provides insights into the technology wage labor market. Class and gender analyses document women's occupation of the worst paid, most tedious and health-destroying segment of the labor market in electronics assembly (OPCS, 1991)—etching circuits onto wafers of silicon, dipping circuits into vats of carcinogenic solvents, and peering through microscopes for seven to ten hours a day to bond wires to silicon chips (Fuentes & Ehrenreich, 1983). Socialist feminist analyses reveal that the extremely low wages paid to women in these jobs, along with women's geographic immobility, may lead sooner to the automation of work done by men, despite its being less menial and more difficult to automate, in order to keep wages low and/or destroy unions (Cockburn, 1981). Socialist feminist approaches also suggest why men dominate the creation of new technologies, since access to venture capital, geographic mobility, and ability to work long hours may be as critical as is technological expertise for the success of start-ups.

USE

Understanding of class relations emerging under capitalism and gender relations under patriarchy helps to explain the intertwining of military and masculinity (MacKenzie & Wajcman, 1999; Enloe, 1983), which drives much technological innovation in the United States and elsewhere. Such understanding also explains the choices made to develop technologies in a certain way—engineering decisions, for example, that favor the wealthy few over relatively less expensive technologies (such as devices for the home) that aid many people, especially women.

Robert Caro's work (1974) revealed that Robert Moses, the master builder of New York's roads, parks, bridges, and other public works from the 1920s to the 1970s, had overpasses built to specifications that discouraged buses on parkways. White upper- and middle-class car owners could use the parkways, such as Wantagh Parkway, for commuting and for accessing recreation sites, including Jones Beach. Because the twelve-foot height of public transit buses prohibited their fitting under the overpasses, blacks and poor people dependent on public transit did not have access to Jones Beach (Winner, 1980).

Current intellectual property rights agreements and laws provide opportunities for choices in technology development that further exacerbate class differences by transferring technologies developed using public moneys to the private realm through patents. The decisions regarding which products are developed falls under the influence of capitalist interests in profit margins. Such intellectual property rights function as a form of privatization (Mohanty, 1997). They allow decisions about which products will be developed to occur in the private, rather than the public, realm. This results in capitalist interests in the bottom line (rather than public needs and interests) dictating which "products" are developed. In the patenting of intellectual property, rights (and profits) get transferred from the public who paid for the research with their tax dollars, to the private company, institution, or individual who controls the patent. Socialist feminists might view this as a transfer from the pockets of the working class, who pay the taxes to underwrite federal research, to the patent holders in the private sector who will reap massive profits, serving the interests of bourgeois capitalists. New technologies in computer science and engineering are often developed using federal grants (paid for by taxes).

DESIGN

Socialist feminism opens the possibilities of more insights into the gender and class distributions within the technology workforce. Understanding that middle- and upper-class men create and design most new technology, along with serving as the sources of money for design and creation, explains much about whose needs are met by current technology and its design. Imagining women as designers, as well as users, of technology suggests that more technologies might meet the needs of women and be adapted for the spaces where women spend time. In the nineteenth century, *New Ideas* magazine (Sept. 1896, p. 214) said, "It is due to the inventive faculty of women that we owe the majority of devices which lighten and perfect labor in their own domain." More recently, Frances Gabe attempted to eliminate the drudgeries of housework by designing a patented house that cleans itself, using 68 separate devices (Macdonald, 1992). Although the self-cleaning house is likely to remain a white, upper-middle-class, suburban solution that would not necessarily improve the lives of lower income women (and might, in fact, eliminate the employment of some), it does hold the potential to improve the lives of many women by reducing housework. In a similar fashion, socialist feminist reform suggests that the allocation of resources for technology development would be determined by greatest benefit for the common good. In a sort of reversal of the Robert Moses overpasses to Jones Beach example, highest priority would go to development of technologies that would redistribute resources to mass transit and provide greater access for the poor.

African American/Womanist and Racial/Ethnic Feminism

Just as socialist feminist theory provided insights into the gender and class distributions of the technology labor market, African American critiques uncover the role of race in the distribution. Racism intertwines and reinforces differing aspects of capitalism and patriarchy.

WORKFORCE

African Americans and Hispanics are underrepresented in engineering and in the upper end of the technology workforce, relative to their percentage in the overall U.S. population (22.4 percent) (NSF, 1999). In 1998, African Americans constituted 3 percent of engineers and 4 percent of computer and mathematical scientists (NSF, 1999, p. 106); 2.5–3 percent of engineers and of mathematical and computer scientists were identified as Hispanic (NSF, 1999, p. 109). Although engineering has been traditionally defined as a career path for mobility from the working to middle class, engineering is pursued by disproportionately fewer blacks and Latinos than whites. Even fewer African American women and Latinas than their male counterparts become engineers or scientists, despite the higher percentage of African American women (compared to African American men) in college (NSF, 1999, p. 113).

In stark contrast, women of color are disproportionately represented in the lowest paying and highest health risk portions of the technology labor force. Studies (Fuentes & Ehrenreich, 1983; Women Working Worldwide, 1991) demonstrate that women of color occupy the ghettos in the cities where the electronic assembly occurs. Outside the technology production workforce, women of color also represent the group most likely to be replaced by technology when automation takes over the work formerly done by their hands. Increasing automation forces women of color from higher paying assembly line factory work into lower paying service sector jobs (Mitter, 1986).

USE

Like socialist-feminism, African American/womanist or racial/ethnic feminism, based on African American critiques of a Eurocentric approach to knowledge, also rejects individualism and positivism for social construction as an approach to knowledge. African American critiques also question methods that distance the observer from the object of study. Because technology, for the most part, involves practical application of more abstract, basic scientific research, the problem of the distance between engineer (researcher) and the technology (object of study) needs to be understood, discussed, and addressed in order to make technological research methodologies clearer to both developers and users. Unlike theoretical scientific projects, which do not necessarily have immediate practical outcomes, projects in computer science or engineering must accommodate the impact of the uses of technology with particular attention to the users of technology. When designing technologies, engineers, designers, and computer scientists would then not only ask how and under what conditions the technology will be used, but they also would have to allow the potential consumer to shape the design of the product. Potentially, the gender, class, and race, along with other factors such as age and ability status, should be considered in defining who the user will be.

DESIGN

A growing recognition has evolved of the strength and necessity for diversity on engineering teams (Thomas J. Joyce, Vice President and Treasurer, Meritor Automotive Inc., personal communication, April 11, 2000). Diversity in gender and race are beginning to be understood as critical, along with the long-established recognition of the importance of having a team representing varied intellectual and technical backgrounds, for designing complex technologies. Because knowledge and consideration of the user/client/customer are central to the technology design, a design team with racial and gender diversity coupled with surveys of demographically diverse customers become instrumental in increasing diversity in technology design.

Since technology generally reflects and reinforces race relations, socialist feminist and African American feminist theories imply that women engineers, computer scientists, and technology designers, through a collective process of political and scientific struggle (Jaggar, 1983), might produce technologies different from those produced by men of any race or class. These technologies might reflect the priorities of more people and be user-accessible and user-friendly for a wider audience because they would be based in the experience of women, whose standpoint as the nondominant group in engineering provides them with a more comprehensive view of reality because of their race, class, and gender. For example, the National Academy of Engineers (NAE) is 97.4 percent male (51 women and 1901 men) and almost exclusively white (Victoria Friedensen, Program Officer, National Academy of Engineering, personal communication, April 13, 2000). One could imagine that if the NAE contained more women and men of color, it might have different priorities and make different recommendations. Women of color in a sexist, racist society are likely to have had very different experiences than members of the NAE. Their experiences and perspectives might in turn lead them to have different priorities and to propose different technologies.

Essentialist Feminism

African American and socialist feminist critiques emphasize race and class as sources of oppression that combine with gender in shaping and being shaped by technology. In contrast, essentialist feminist theory posits that all women are united by their biology. Women are also different from men because of their biology, specifically their secondary sex characteristics and their reproductive systems. Frequently, essentialist feminism may extend to include gender differences in visuo-spatial and verbal ability, aggression and other behaviors, and physical and mental traits based on prenatal or pubertal hormone exposure. For example, some sociobiological research (Wilson, 1975; Trivers, 1972; Dawkins, 1976) and some hormone and brain lateralization research seems to provide biological evidence for differences in mental and behavioral characteristics in males and females. Camilla Benbow and Julian Stanley's work (1980) hypothesizing X-linked genes for visuo-spatial ability provided a biological rationale for the larger numbers of males with superior performance on tests of mathematical (especially visuo-spatial) ability. These differences often serve as essentialist arguments for the absence or exclusion of women from engineering and technology. For example, visuo-spatial abilities remain as the subfield which accounts for the approximately 35–point average gender difference between males and females on the math SAT. Visuo-spatial abilities and skills become a critical filter for engineering because of their perceived importance for design.

WORKFORCE

Biological bases for sex-based differences in aggression have been studied in other species (Trivers, 1972; Wilson, 1975), lower primates (Hrdy, 1986; Yerkes, 1943), and human children (Maccoby & Jacklin, 1974), and adult humans (Bleier, 1979; 1986). Usually the higher aggression levels in males have been attributed to exposure to higher levels of testosterone in postpuberty and/or prenatally. The behavioral variable of aggression becomes difficult to disentangle from the average larger physical size, increased muscle mass, lengthened vocal cords, and other secondary sexual characteristics resulting from testosterone in male humans that may contribute to, and be perceived as, aggression. Aggression also becomes entangled with the characteristic of an engineer and the development of technology. Not only do others cite the competitive nature of engineering (Cockburn, 1983; Wacjman, 1991) and computer science (AAUW, 2000) as reasons women have not entered these fields in large numbers despite three decades of support for women in engineering programs, but women themselves also cite that reason:

> Female students said they were turned off by violent software games and felt the computer world is dominated by adolescent males. . . . When asked to describe a person who was really good with computers, they described a man.

In a 1997 survey of 652 college-bound high school students in Silicon Valley, Boston and Austin, Texas, 50 percent of all students said the field of computer science was "geared toward men." (Knight-Ridder, 2000)

Testosterone and aggression define masculinity in our culture and most others. Cockburn (1983), Hacker (1989), and others (Wacjman, 1991) have explored the co-definitions and reinforcement of the cultures of masculinity and technology, including the technologies of IT (Hacker, 1981, 1989). The military and war become an obvious site where aggression, technology, and masculinity conjoin. The historical origins of much technology in the military, where aggression and masculinity become defined and valued, remain, despite some deployment of women and use of nonaggressive approaches in modern defense. Historically and currently, war serves as the ultimate site of contest, used for a variety of nationalist purposes.

USE

An essentialist feminist approach suggests that men, because of their biology and inability to conceive, develop technologies to dominate, control, and exploit the natural world, women, and other peoples (Easlea, 1983). Women, in contrast, because of their biology, not only have less testosterone, but also have the ability to give birth. Giving birth gives women less direct control over their bodies and connects them more closely with nature, other animals, and life (Merchant, 1979; King, 1989). In its most simplistic extreme form, essentialism implies that men use technologies to bring death and control to other people, women, and the environment, while (or because) women give birth and nurture life in all its forms. In his study of the discovery and development of nuclear weapons and the atomic bomb, Brian Easlea (1983) examines the language and behavior of the scientists. Analyzing the aggressive sexual and birth metaphors the scientists use to describe their work, he argues that men "give birth" to science and weapons to compensate for their inability to give birth to babies.

DESIGN

Webster calls this essentialist feminist critique "eco-feminism." Eco-feminism represents one type of essentialist feminism, usually associated with the left-wing political leanings that Webster implies in her discussion of ecofeminism and technology (Webster, 1995). Essentialism can be used to support either superiority or inferiority of women compared to men, as long as the source of difference remains rooted in biology. Feminist scientists

(Bleier, 1979; Fausto-Sterling, 1992; Hubbard, 1979; Rosser, 1982) critiqued essentialism and the sociobiology research supporting it as not providing sufficient biological evidence to justify differences in mental and behavioral characteristics between males and females. Essentialism was seen as a tool for conservatives who wished to keep women in the home and out of the workplace. Eventually, feminists re-examined essentialism from perspectives ranging from conservative to radical (Corea, 1985; Dworkin, 1983; MacKinnon, 1982, 1987; O'Brien, 1981; Rich, 1976) with a recognition that biologically based differences between the sexes might imply superiority and power for women in some arenas.

> Girls said they use computers to communicate and perform specific tasks, while boys have underdeveloped social skills and use computers to play games and "fool around." Turkle said: "Instead of trying to make girls fit into the existing computer culture, the computer culture must become more inviting for girls." The report said girls and women cannot settle for being consumers of technology. They must be prepared to become designers and creators if they are going to fully participate and shape the new computer age. (AAUW, 2000)

Both eco-feminism and essentialism suggest that because of their biology, women would design different technologies and use them differently. Indeed, the studies done of inventions by women and surveys of patents obtained by women (Macdonald, 1992) suggest that many women develop technologies related to reproduction (e.g., Nystatin to prevent vaginal yeast infections), secondary sex characteristics (backless bra), or babies/children (folding crib). An essentialist feminist theoretical approach to these invention and patent data studies implies that differences in women's, compared to men's, biology—differences such as hormone levels, menstruation, giving birth, and ability to lactate to nourish offspring—leads to women designing different technologies and using technologies differently from men.

Existentialist Feminism

In contrast to essentialist feminism, existentialist feminism purports that it is not the biological differences themselves, but the value that society assigns to biological differences between males and females that has led women to play the role of "the Other" (Tong, 1989). The philosophical origins of existentialist feminism emphasize that it is man's conception of woman as "other" that has led to his willingness to dominate and exploit her (Beauvoir, 1948).

WORKFORCE

The gender stratification of the technology labor market, where men occupy the higher-paying roles of engineer and designer and women occupy the lower-paying positions of worker, is a value assignment based on biological differences between the sexes. As Valerie Frissen's (1995) work demonstrates, women do the casual work of word processing and telecommuting in their home; this lower-paid work fits with women's role as other and the value assigned to her role in childcare by society.

DESIGN

Existentialist feminism would explain the gender polarization of technology, where men design technology and women use it, as an example of his domination of her as other. As Webster (1995, p. 149) points out, "the 'objectivity' which is aspired to in the design process is the viewpoint of the men who design." Simone de Beauvoir described the phenomenon as follows: "Representation of the world, like the world itself, is the work of men; they describe it from their own point of view, which they confuse with the absolute truth" (Beauvoir, 1948).

An existentialist feminism framework might be used to explain the higher frequency of inventions by women of technologies useful for menstruation, childbirth, lactation, and hormones. In contrast to essentialism, rather than placing the emphasis for the origin of the technology on the biology itself, existentialism would suggest that it is value assigned by society to women as other that leads to the technology. Women serve as the predominant caretakers of babies and children, perhaps because they give birth to them and nurse them. Existentialist feminism would suggest that this assignment of the role as other based on the biological reasons would lead to women having more experience caring for babies and children. In turn, this experience would lead them to invent more technologies useful for childcare, such as the pull-down-from-the-wall baby-changing stations found in public restrooms, disposable diapers, and portable folding cribs (Macdonald, 1992).

In similar fashion, the computing industry in its language, as well as in its hardware, reflects the world of the designers. That world reflects the world of men who spend less time in childcare and have relatively more time to "crack the code."

> And once you've done that [spent hours in front of screens] ... you'd know there were things you could exchange, a level of hints and tips, but more deeply a level of understanding, shared language ... but they [those men]

don't look around and see the absence of women, they don't perhaps think that they are creating a language, and perhaps, as it's expressed at work, a power-base which is exclusive. And they don't realise that they are where they are because the computing industry is designed for them. It's designed for people who *have* got hours to spare: in the garage, or the shed, or the attic, cracking the code. (Male library research officer, quoted in E. Green, J. Owen, & D. Pain, 1993, pp. 145–146)

USE

In even more direct ways, the use of the male norm excludes women as users of the technology. Military regulations often apply Military Standard 1472 of anthropometric data so that systems dimensions use the 95th and 5th percentile of male dimensions in designing weapons systems. This led to the cockpits of airplanes being designed to fit the dimensions of 90 percent of male military recruits (Weber, 1997). This worked relatively well as long as the military was entirely male. In the case of the Joint Primary Aircraft Training System (JPATS), used by both the Navy and Air Force to train its pilots, the application of the standard accommodated the 5th through 95th percentile (90 precent) of males, but only approximately the 65th through 95th percentile (30 percent) of females. The policy decision by Secretary of Defense Les Aspin (1993, p. 1) to increase the percentage of women pilots uncovered the gender bias in the cockpit design. Designed to exclude only 10 percent of male recruits by its dimensions, the cockpit excluded 70 percent of women recruits, making it extremely difficult to meet the military's policy goal of increasing the number of women pilots. The officers initially reacted by assuming that the technology reflected the best or only design possible and that the goal for the percentage of women pilots would have to be lowered and/or the number of tall women recruits would have to be increased. This initial reaction, which represented the world viewpoint of men (Beauvoir, 1948), changed. When political coalitions, the Tailhook scandal, and feminist groups reinforced the policy goal, a new cockpit design emerged which reduced the minimum sitting height from 34 to 32.8 inches, thereby increasing the percentage of eligible women (Weber, 1997, p. 379).

Psychoanalytic Feminism

Based on the Freudian notion that anatomy is destiny, psychoanalytic theory assumes that biological sex will lead to different ways for boys and girls to resolve the Oedipus and castration complexes that arise during the phallic stage of normal sexual development. Rejecting the biological determinism in Freud, Dorothy Dinnerstein (1977) and Nancy Chodorow (1978), in particular, have used an aspect of psychoanalytic theory known as object relations theory to examine the construction of gender and sexuality during the Oedipal stage of psychosexual development, which usually results in male dominance. They conclude that the gender differences resulting in male dominance can be traced to the fact that in our society, women are the primary caretakers of most infants and children.

WORKFORCE

Evelyn Keller (1982, 1985) in particular applied the work of Chodorow and Dinnerstein to suggest how science, populated mostly by men, has become a masculine province, in its choice of experimental topics; use of male subjects for experimentation, interpretation and theorizing from data; as well as the practice and applications of science undertaken by the scientists. Keller suggests (1982, 1985) that since the scientific method stresses objectivity, rationality, distance, and autonomy of the observer from the object of study (i.e., the positivist neutral observer), individuals who feel comfortable with independence, autonomy, and distance will be most likely to become scientists. Feminists have suggested that the objectivity and rationality of science are synonymous with a male approach to the physical, natural world.

DESIGN

A psychoanalytic feminist framework might provide the theoretical backdrop for Cockburn's (1981, 1983, 1985) work documenting the intertwining of masculinity and technology. Encouraged to be independent, autonomous, and distant, male engineers and computer scientists design technologies and IT systems reflecting those characteristics. As Suzanne Bodker and Joan Greenbaum (1993) suggest, the "hard-systems" approach to computer systems development follows the positivist, linear, and technicist approach compatible with Western scientific thought. The technical capabilities, constraints of the machines, and rational data flow become the focus and driver of the technology design.

This hard-systems design approach used by (mostly male) developers of computer systems assumes separation, distance, and independence on different levels: 1) between the abstract systems development and the concrete real world of work: separation ignores the often circular and interconnected forces of organization, assuming that they remain linear and unaffected by other hierarchical, power relations; 2) between the developers and users:

since users do not contribute to the design of the system, their needs and suggestions, which might make the system function more smoothly in the real world of work, are ignored. The problems caused by this abstraction, objectivity, autonomy, and separation have spawned new methods such as "soft-systems" human factors approaches to solving the problems and mediating the gap.

USE

The gender constellation predicted by psychoanalytic feminism becomes transparent: the men who design hardware systems design them in ways reflective of their perspective on the world with which they feel comfortable. Such system designs tend to place priority on data and ignore relationships between people. Women, socialized to value connections and relationships, tend to feel uncomfortable with the hard-systems approach. As users, they find that the technology fails to aid much of the real-world work. The design inhibits or fails to foster good teamwork and other relationships among co-workers. Because the design does not reflect their view of priorities in the organization and work, and actively ignores the reality of power and gender relations, women tend to be excluded, and exclude themselves, from hard-systems design.

Critiques from a psychoanalytic feminist perspective raise the very interesting question of how systems design might change if more feminine values and connection became priorities. Knut Sorenson (1992) explored whether male and female computer scientists work differently. He found that men tended to focus on mathematical models and computer programming while women spent more time running experiments, reading scientific literature, and plotting data. After studying the technological and political values of men and women engineering students, graduate students, and junior R&D scientists at the Norwegian Institute of Technology, Sorenson found that women brought "caring values" to research in computer science. "Caring values" included empathy and rationale of responsibility. "In computer science, this means that women have a caring, other-oriented relationship to nature and to people, an integrated, more holistic and less hierarchical worldview, a less competitive way of relating to colleagues and a greater affinity to users" (Sorenson, 1992, p. 10).

Understanding the importance of relationships and power, some women computer scientist designers (Suchman, 1994; Microsyster, 1988) have attempted to link users with systems design as an explicit attempt to empower women. Although some might view this as an example of Sandra Harding's "strong objectivity," this shortening of distance between the user and the system design mimics Keller's description of Barbara McClintock's work in *A Feeling for the Organism*. In the shortening of the distance between the observer and the object of study, Keller describes less autonomy, independence, and separation as classic hallmarks of psychoanalytic feminism when applied to the work of women scientists.

Radical Feminism

Radical feminism maintains that women's oppression is the first, most widespread, and deepest oppression (Jaggar & Rothenberg, 1992). Scientific institutions, practice, and knowledge are particularly male-dominated and have been documented by many feminists (Bleier, 1984; Fee, 1983; Griffin, 1978; Haraway, 1978, 1989; Hubbard, 1990; Keller, 1985; Merchant, 1979; Rosser, 1990) to be especially effective patriarchal tools to control and harm women. Radical feminism rejects most scientific theories, data, and experiments precisely because they not only exclude women but also because they are not women-centered.

DESIGN

Some might define the work of Tone Bratteteig and her co-workers as radical feminism because it originates from women's discourse on computer science problems and methods. Indeed, they insist on prioritizing applicability of systems and putting users and developers in the same plane as collaborators in systems development. This concept of starting from the understanding of a woman worker and her abilities and then focusing on how her professional competence can be augmented by the use of a system does begin with women's experience. This focus on women (Thoresen, 1989; Hacker, 1990) is consistent with feminist principles.

In addition to the focus on women and seeking to empower women, Catharine MacKinnon (1987) adds a further criterion to radical feminism. She suggests that the consciousness-raising group provides a methodology for radical feminism. Because patriarchy pervades and dominates all institutions, ideologies, and technologies, women have difficulty placing their experiences, lives, and needs in central focus in everyday life and environments. A successful strategy that women use to obtain reliable knowledge and to correct patriarchal ideology is the consciousness-raising group (Jaggar, 1983). Using their personal experiences as a basis, women meet together in communal, nonhierarchical groups to examine their experiences in order to determine what counts as knowledge (MacKinnon, 1987). Lucy Suchman, in her work at Xerox's Palo Alto Research Center (PARC), uses this approach to further her view that knowledge held by users is central to the design process. They are "taking computerization as an occasion to articulate unacknowledged forms of expertise and to take that knowledge seriously as a basis for design" (Suchman & Jordan, 1989, p. 158). In recognizing the gap between technologies created by experts and their use in real working environments, they have produced a "radical reconceptualisation of the computer systems development process which recognises the innovatory character of the implementation process and places the local expertise of users at the centre" (Webster, 1995, p. 165).

The interplay between designing and use, or designer and user, described in the Suchman example, may also be seen to demonstrate the cyclicity often associated with radical feminism: because radical feminists believe in connection and a conception of the world as an organic whole, they reject dualistic, hierarchical approaches and dichotomies that fragment the organic whole of reality. Cyclicity as a conception of time, and thinking as an upward spiral, seem more appropriate ways to study a world where everything is connected in a process of constant change (Daly, 1978, 1984).

WORKFORCE

For example, radical feminists might interpret the binary 0,1 foundation of computers and computing as being based on the primary dichotomy/dualism of male–female. The "switchers," "controls," and "operations" language of computing fit the patriarchal mode of control. The dichotomy receives reinforcement by the "domination of men and the absence of women from the design process. [It] is *one* factor which creates technologies which are closely geared to the needs of men and which are inappropriate to women's requirements" (Webster, 1995, p. 179).

To understand the complete, comprehensive influence of patriarchy, and to begin to imagine alternative technologies, lesbian separatism would suggest that women must separate entirely from men. Lesbian separatism, often seen as an offshoot of radical feminism, would suggest that separation from men is necessary in a patriarchal society for females to understand their experiences and explore the potential of science and the impact of technologies. Cockburn advocates women-only organization:

> In my view, by far the most effective principle evolved to date is separate, woman-only organisation. It enables us to learn (teach each other) without

being put down. Provide schoolgirls with separate facilities and the boys won't be able to grab the computer and bully the girls off the console. Provide young women with all-women courses so that they can gain the experience to make an informed choice about an engineering career. We need to demand a massive increase in resources from the state, from industry, from industrial training boards, for women-run, women-only, initiatives. Everywhere we have tried it, from women's caucuses to Greenham Common (the women's peace camp at a cruise missile base), autonomy works wonders for our feelings and our strength. We need, before all else, a great expansion of the autonomous sphere in technology. (Cockburn, 1983, p. 132)

The establishment of engineering at Smith College, a women's college, may provide a site where ideas, curriculum, and pedagogy in technology can be explored in an environment somewhat separate from men.

USE

Radical feminism would suggest that the reason no truly feminist alternative to technology exists is that men, masculinity, and patriarchy have become completely intertwined with technology and computer systems in our society. Imagining technology from a woman-centered perspective in the absence of patriarchy becomes extremely difficult, if not impossible. Since engineering and technology development in the West/North foreground control—control over nature, over people, and over machines—imagining a technology premised on cooperation, collaboration, and working with nature, people, and machines runs contrary to our image of the technology that evolved in a patriarchal, heterosexist society. Scandinavians (Gunnarsson & Trojer, 1994) suggest that the creation and protection of human life should be the point of departure for technological development for women:

> Women's ethics . . . is not sentimental. It is practical. It implies a concrete and holistic consideration of people's need for a sustainable environment and that basic security which is the precondition of common responsible action. . . . A step by step process . . . makes the protection of the weak its highest priority: creating social solidarity and collective security in a sustainable manner from below. This must go hand in hand with a gradual reduction of the importance of wage labour in the shape of a collective re-appropriation of control over the means of subsistence. . . . In the context of such a process new technology must be invented, old technologies transformed or abandoned. (p. 79)

Queer and transgender theories question links between sex, gender, and sexual orientation (Butler, 1990). They raise additional challenges about the links between economic, racial, and dominance factors with gender in our society. As Judith Butler argues (1990, 1992, 1994), the very act of defining a gender identity excludes or devalues some bodies and practices, while simultaneously obscuring the constructed character of gender identity; describing gender identity creates a norm. The creators of "The Turing Game" a computer game modification of Alan Turing's suggestion of ways to differentiate machines from people and men from women, explain their goals and methodologies in the following terms:

> Do men and women behave differently online? Can you tell who is a man and who is a woman based on how they communicate and interact with others on the Internet? Can you tell how old someone is, or determine their race or national origin? In the online world as in the real world, these issues of personal identity affect how we relate to others. Societies are created and destroyed by these understandings and misunderstandings in the real world. Yet, as the online world becomes increasingly a part of our lives, identity in this new medium is still poorly understood. At the Georgia Institute of Technology, we have created an online game to help us explore and teach about these issues. This environment, called the Turing Game, is a game of identity deception, expression, and discovery. Available on the Internet, it has been played by more than 9,000 people. Players from seventy-six countries on all seven continents have used the game to learn about issues of identity and diversity online through direct experience. At the same time, they have created communities of their own, and explore the boundaries of electronic communication. (Berman & Bruckman, 2000)

This "Turing Game" explores the creations of these norms and how the Internet opens possibilities for identity changes and deception (see http://www .cc.gatech.edu/elc/turing).

Postmodern Feminism

Liberal feminism suggests that women have a unified voice and can be universally addressed (Gunew, 1990). According to postmodernism, "the values of reason, progress, and human rights endorsed by the Enlightenment have shown their dark side" (Tanesini, 1999, p. 239). In postmodernism, the self is no longer regarded as "masterful, universal, integrated, autonomous, and self-constructed; rather, it is socially constructed by ideology, discourse, the structure of the unconscious, and/or language" (Rothfield, 1990, p. 132). Postmodernism dissolves the universal subject and postmodern feminism dissolves the possibility that women speak in a unified voice or that they can be universally addressed. Postmodern perspectives stress that due to her situatedness—the result of her specific national, class, and cultural identities—the category of woman can no longer be regarded as smooth, uniform, and homogeneous. At least some postmodern feminists (e.g., Cixous & Clement, 1986; Kristeva, 1984, 1987) suggest that women, having been marginalized by a dominant male discourse, may be in a privileged position, that of outsider to the discourse, to find the holes in what appears solid, sure, and unified. Otherwise, the dominant discourse threatens to rigidify all in society along previously established lines. Postmodern feminist theories imply that no universal research agenda or application of technologies will be appropriate and that various women will have different reactions to technologies depending upon their own class, race, sexuality, country, and other factors.

This definition of postmodern feminism parallels the description of the complex and diverse co-evolution of women and technology, particularly computing technology. Many gender and technology studies have classically focused on the exclusion of women from technological design, from uses most appropriate to their needs, and from higher-paying and decision-making jobs in the technology workforce. These studies tend to imply a universalist stance that all women have similar needs for uses of technology and that the employment categories and effects within technology industries affect women uniformly.

DESIGN

As postmodern feminist theory recognizes limitations of perceiving women as a universal group, so have deeper, more complex studies of technology industries revealed limitations of simplistic assumptions in technology designs. For example, Webster suggests that critiquing the absence of understanding of women's needs is easier than identifying what women's needs and priorities really are.

> First of all, they are of course not uniform across all women or across all social classes, nationalities and cultures. Then, making the link between women's needs and the consequently desirable features of an IT system is a vexed question, and one that is too often dealt with in an essentialist manner by reference to the supposedly universal characteristics of women IT users (for example, as lacking in confidence with computers, as disinterested in computers except as tools, as caring and sociable rather than rational and technical). Feminist design initiatives have often encountered this difficulty. Most have

addressed it by focusing away from the eventual characteristics of the IT *products* which they shape, and towards the *processes* of shaping which offer scope for permanently redefining the role of female users in systems design. (Webster, 1995, p. 178)

WORKFORCE

Just as women's needs for IT or technology designs differ and vary, depending upon class, nationality, culture, age, and other factors, employment of women in technology industries also does not fit a universal or uniform pattern. Innovations in technology have not led to overall restructuring of established sexual divisions of labor, or unequal gender or race relations (Kirkup, 1992, p. 81). Some groups of women have gained or lost ground in their employment in technology industries. For example, some women have benefited from programs designed to increase female representation in IT and other technology industries. These equity and access programs (based in liberal feminist theories) have benefited some professional middle-class women whose educational backgrounds position them to capitalize on better employment opportunities (Wickham & Murray, 1987).

Although relocation and temporization of work have tended to hurt employees in general and women in particular, the effects may depend on urban location. For example, closing offices in city centers and metropolitan areas has tended to hurt urban women, who are more likely to be of lower socioeconomic status and of color, while creating employment for women in the suburbs (Greenbaum, 1995). In contrast, development of offshore information processing has improved employment for women in poorer countries. Information and data processing functions, once performed by women in the First World have now been exported to low-cost economies because telecommunications and satellite technologies make this possible (Webster, 1995, p. 182).

USE

Similarly, the "flexibility" and "casualization" of the workforce, which telecommuting permits, may hurt wages, benefits, and long-term stability overall. Although it creates and/or increases the double burden for women who can mind children while working at home, some women prefer this option to no work at all.

> Women have always needed to find ways of managing their domestic responsibilities whilst in employment, and information communication technology (ICT)-supported forms of working, such as teleworking, as well as other

"contingent" forms of employment, such as part-time work, offer them the means to do this. To the extent that ICTs provide the wherewithal for carrying these casual forms of employment to new heights of sophistication, they are an important mechanism in the confirmation of women's marginal relationship to the labour market.

But ICTs can also directly allow women to handle paid and unpaid work simultaneously. A simple example is the cell phone. Frissen (1995) suggests that it allows working women to transform the double burden of home and work into a parallel burden, allowing them to multitask and manage their domestic responsibilities better, from a distance. (Webster, 1995, pp. 184–185)

This quotation suggests why women may react differently to technologies, depending upon their race, class, age, ability status, parental status, urban or rural location, or other factors. Coupled with the rapid and changing pace of technology, postmodern feminism suggests why universal theories fail to fit the reality of women's lives. The lack of universalism may inhibit gender-based coalitions and organizing, making it also easier to understand the political inactivism, of which individuals who articulate postmodern perspectives may be accused (Butler, 1992).

Postcolonial Feminism

Beginning in 1947, following various campaigns of anticolonial resistance (often with an explicitly nationalist basis), many colonial empires formally dissolved, and previously colonized countries gained independence (Williams & Crissman, 1994). Although the end of colonial rule created high hope for a proper postcolonial era, the extent to which the West had not relinquished control became clear quickly. The continuing Western influence—particularly in the economic arena, but also in the political, ideological, and military sectors—became known as neocolonialism by Marxists (William & Chrisman, 1994). Feminists have suggested that patriarchy dominates everyday life in postcolonial and neocolonial periods, much as it dominated colonial life.

Not surprisingly, technologies reflect the varying complex aspects of the interrelationships among developed and developing countries in general and between the particular cultures of the colonized and colonizing countries. General themes include the underdevelopment of the southern continents by Europe and other northern continents (Harding, 1998); ignoring, obscuring, or misappropriating earlier scientific achievements and history of countries in southern continents; the fascination with so-called "indigenous science"

(Harding, 1998); the idea that the culture, science, and technology of the colonizer or former colonizing country remains superior to that of the colony or postcolonial country; and the insistence that developing countries must restructure their local economies to become scientifically and technologically literate in order to join and compete in a global economy (Mohanty, 1997). In northern, former colonizing countries, the concurrent restructuring effects of multinational corporations and other forces of globalization are evidenced in downsizing, privatization, and widening economic gaps between the poor and the very wealthy. The particular forms and ways that these general themes take shape and play out varies, depending upon the history, culture, geography, and length of colonization for both the colonized and colonizing countries.

WORKFORCE

As suggested by some of the examples in the section on postmodernism, many women in so-called "Third World" or developing countries receive employment in technology industries or because of technological developments such as satellites that permit rapid data transmission over large geographic distances. The United States, Western Europe, and Japan house the corporate headquarters, owners, and decision makers of these global, multinational corporations; technological developments permit these companies to roam the globe and use women in offshore, formerly colonized, and/or developing nations as cheap sources of labor. Because new technologies transcend boundaries of time and space, they facilitate corporations in dispersing work around the globe to exploit sexual and racial divisions of labor.

In some ways the clothing industry led the way in using technologies to take advantage of lower labor costs. During the 1960s and 1970s, European, American, and Japanese companies located plants in Asia and the Far East to reduce labor costs (Elson, 1989). More recently, subcontracting (rather than owning plants) has permitted clothing companies to derive benefits of women's labor in developing countries for the garment industry without employing the women directly. Peter Dicken (1992) documents that German firms buy from offshore firms in Eastern Europe and Magreb countries; the United States from Mexico, the Philippines, and Caribbean; Britain from Portugal, Morocco, and Tunisia. These arrangements permit the design and cutting operations to be performed by men earning higher wages using skilled technology in the developed countries, while women receive low wages to sew and assemble garments in offshore, subcontracted, small companies in the developing world.

USE

Information technology, satellites, and computerization become the glue that holds the global networks within a company together and permits them to function smoothly and efficiently. Benetton Group exemplifies this:

> At [Benetton's] Italian headquarters is a computer that is linked to an electronic cash register in every Benetton shop; those which are far away, like Tokyo and Washington, are linked via satellite. Every outlet transmits detailed information on sales daily, and production is continuously and flexibly adjusted to meet the preferences revealed in the market. (Elson, 1989, p. 103)

DESIGN

The IT industry itself uses subcontracted female labor in developing countries, particularly for software development. As Heeks (1993) points out, Western managements control the conduct of software development projects, relying on women from India, China, Mexico, Hungary, and Israel to use as programmers. Telecommunications technologies ease the transmission of specifications and completed work between the workers in developing countries and client companies in the West. Women from these developing countries are preferred over workers in developed countries because of their technical and English proficiency, relatively high productivity, and relatively low labor costs.

The same factors of technical and language proficiency and low-wage costs, coupled with satellite and telecommunications systems, have transformed the clerical industry. The financial services sector, in particular, has capitalized on computerization of clerical functions and the use of satellite and telecommunications functions to tap large pools of female clerical labor in Mexico, the Caribbean, southeast Asia, China, India, and Ireland (Probert & Hack, n.d.).

These examples from the garment, software, and clerical services industries clearly demonstrate aspects of postcolonialism inasmuch as control of the economy in developing countries remains in the hands of developed countries. And it is patriarchal control, since women, not men, in the developing countries become the sources of cheap labor. Language becomes an interesting feature that continues to tie former colony with colonizer. Theoretically, satellites and telecommunications transcend geographical barriers and permit any developed country to use labor in any developing country. Practically, the former ties developed between colony and colonizer, as well as the language of the colonizer learned by the colonized, means that former relationships continue in the neocolonial, modern world. Does the conversion of some Indian universities to software factories exemplify this language connection in the IT world where English dominates?

Cyberfeminism

Cyberfeminism stands not only as the most recent feminist theory but also as the theory that overtly fuses technology with gender. As the name suggests, cyberfeminism explores the ways that information technologies and the Internet provide avenues to liberate (or oppress) women. In the early 1990s, the term "cyberfeminism" gained use in various parts of the world (Hawthorne & Klein, 1999), with VNS Matrix, an Australian-based group of media artists being one of the first groups originating the term.

DESIGN

The individuals who defined cyberfeminism (Plant, 1996; Millar, 1998) saw the potential of the Internet and computer science as technologies to level the playing field and open new avenues for job opportunities and creativity for women: "A woman-centered perspective . . . advocates women's use of new information and communications technologies of empowerment. Some cyberfeminists see these technologies as inherently liberatory and argue that their development will lead to an end to male superiority because women are uniquely suited to life in the digital age" (Millar, 1998, p. 200). Absence of sexism, racism, and other oppression would serve as major contrasts between the virtual world and the real world: "Cyberfeminism as a philosophy has the potential to create a poetic, passionate, political identity and unity without relying on a logic and language of exclusion. It offers a route for reconstructing feminist politics through theory and practice with a focus on the implications of new technology rather than on factors which are divisive" (Paterson, 1994).

WORKFORCE

The early days of computer science and information technologies seemed to reflect these dreams. In 1980, women represented 37 percent of computer science majors. As the series of articles revisiting the early history of computing revealed, Ada Lovelace contributed to the development of the protocomputer, and Grace Hopper invented virtual storage and created the first computer language composed of words (Stanley, 1995). Women performed calculations and wired hardware for the first digital electronic computer, ENIAC (Electronic Numerical Integrator and Computer) (Perrolle, 1987). In the late 1980s, however, a drastic change began to occur. Numbers of women majoring in computer science plummeted; this plunge coincided with the restructuring of the capitalist system on a global scale and with the rise of financial speculation permitted by the nonproductive economic investment in the new information technologies (Millar, 1998). As Melanie Millar and other cyberfeminist critics point out, the existing elites have struggled to seize control and stabilize the commercial potential of digital technologies, as well as their research and development. Discontinuity, speed, symbolic and linguistic spectacle, and constant change characterize information technology and digital discourse. Although these characteristics of instability and indeterminacy caused by changing technology open the possibility for other changes in the social realm and within power relations, it is very unclear that information technologies and cyberculture will result in such social changes.

USE

Some critics suggest that the current information technology revolution has resulted in a rigidifying and reifying of current power relations along previously existing gender, race, and class lines. "When I did a net search of Filipina women, all that came up [were] thumbnail pictures of Filipina women who 'like American men.' I was angry and disgusted." (Pattanaik, 1997). The Internet becomes a tool making women more vulnerable to men using it for ordering brides from developing countries, prostitution, cybersex, assumption of false identities, and pornography (Hawthorne & Klein, 1999).

Despite their postmodern veneer of fragmentation, shifting identities, and speed, information technologies rest upon the power of science and technology to emancipate humans and rest on a faith in abstract reason. Millar (1998) defines this situation as "hypermodern," describing the packaging of modern power relations that are universally patriarchal, racist, and bourgeois in a postmodern discourse of discontinuity, spectacle, and speed.

This raises the question of whether cyberfeminism is really a feminist theory. Could cyberfeminism merely represent an attempt to see information technology as the latest venue for women's liberation, much as Shulamith Firestone (1971) envisioned such liberation resulting from reproductive technologies? Although reproductive technologies have resulted in significant feminist critiques, theorizing, and discussion, no one considers them to constitute a feminist theory.

Each of several feminist theoretical perspectives used here to examine the relationships among women, gender, and technology might also be applied to the Internet and information technologies or aspects of what is being called cyberfeminism. Liberal feminism would raise issues of women's equal access to digital hardware and the digital world. Socialist feminism would ask questions about class in terms of who profits from and who loses from cybertechnology. African American or racial/ethnic feminism would focus on issues of race and the "Digital Divide." Essentialist feminism would look at biologically based differences, whereas existentialist feminism and psychoanalytic feminism might raise questions of gender difference based on women's role as caregiver and concerns about socialization (how to remain connected with friends, workers, and families). Radical feminists would point out the potential use of IT to create alternative versions of reality and even to avoid interactions with men in the real world. Postmodern feminists would underline the fast, fragmented, and situation-dependence of information technologies and digital discourse and the impossibility of universalizing its use, design, or impact for all women. Postcolonial feminists might highlight the widening gap between the haves—who may be software designers or members of one of the universities now being converted to specialize in educating information technology workers—and the have-nots, who do not even have electricity, let alone computers, in their homes, as the globalization of the IT industry continues. Although cyberfeminism fuses technology, women, and feminism, it appears not to be a theory; instead, cyberfeminism itself can be critiqued using other feminist theories.

Taken together, the spectrum of feminist theories provides different, new insights to explore both gender and technology. Since many feminist theories emerged in response to critiques of one or more preceding theories, successor theories tend to be more comprehensive and/or compensatory for factors or groups overlooked by previous theories. Expanding the theoretical perspectives beyond liberal, ecofeminist, and even socialist feminism to embrace other feminist theoretical perspectives provides corrections and extensions to the co-evolution of gender and technology.

References

Abbate, Janet. (1999). Cold war and white heat: The origins and meanings of packet switching. In Donald MacKenzie & Judy Wacjman (Eds.), *The social shaping of technology* (2nd ed., pp. 351–379). Philadelphia: Open University Press.

Aspin, Les. (1993, April 28). *Policy on the assignment of women in the armed forces.* Washington, D.C.: Department of Defense.

Association of American University Women. (2000). *Tech-savvy: Educating girls in the new computer age.* Washington, D.C.: AAUW Educational Foundation.

Barnaby, F. (1981). Social and economic reverberations of military research. *Impact of Science on Society, 31*, 73–83.

Beauvoir, Simone de. (1947). *The second sex* (H. M. Parshley, Trans. & Ed.). New York: Vintage Books.

Benbow, Camilla, & Stanley, Julian. (1980). Sex differences in mathematical ability: Fact or artifact? *Science, 210*, 1262–1264.

Berg, Anne-Jorunn. (1999). A gendered socio-technical construction: The smart house. In Donald MacKenzie & Judy Wacjman (Eds.), *The social shaping of technology* (2nd ed., pp. 301–313). Philadelphia: Open University Press.

Berman, Joshua, & Bruckman, Amy. (2000). The Turing Game: A participatory exploration of identity in online environments. In *Proceedings of directions and implications of advanced computing (DiAC) 2000*. Seattle, Wash.: Computer Professionals for Social Responsibility.

Birke, Lynda. (1986). *Women, feminism, and biology: The feminist challenge*. New York: Methuen.

Bleier, Ruth. (1979). Social and political bias in science: An examination of animal studies and their generalizations to human behavior and evolution. In Ruth Hubbard & Marian Lowe (Eds.), *Genes and gender II: Pitfalls in research on sex and gender* (pp. 49–70). New York: Gordian Press.

———. (1984). *Science and gender. A critique of biology and its theories on women*. New York: Pergamon Press.

———. (1986). Sex differences research: Science or belief? In Ruth Bleier (Ed.), *Feminist approaches to science* (pp. 147–164). New York: Pergamon Press.

Bodker, S., & Greenbaum, J. (1993). Design of information systems: Things versus people. In E. Green, J. Owen, & D. Pain (Eds.), *Gendered by design: Information technology and office systems* (pp. 53–63). London: Taylor and Francis.

Butler, Judith. (1990). *Gender trouble: Feminism and the subversion of identity*. New York: Routledge.

———. (1992). Introduction. In J. Butler & J. Scott (Eds.), *Feminists theorize the political* (pp. xii–xvii). New York: Routledge.

———. (1994). *Bodies that matter: On the discursive limits of "sex."* New York: Routledge.

Caro, Robert. (1974). *The power broker: Robert Moses and the fall of New York*. New York: Random House.

Chodorow, Nancy. (1978). *The reproduction of mothering: Psychoanalysis and the sociology of gender*. Berkeley: University of California Press.

Cixous, Helene, & Clement, C. (1986). *The newly born woman*. Minneapolis: University of Minnesota Press.

Cockburn, Cynthia. (1981). The material of male power. *Feminist Review, 9*, 41–58.

———. (1983). *Brothers: Male dominance and technological change*. London: Pluto Press.

————. (1985). *Machinery of dominance: Women, men and technical know-how.* London: Pluto Press.

Corea, Gena. (1985). *The mother machine: Reproductive technologies from artificial insemination to artificial wombs.* New York: Harper & Row.

Cowan, R. S. (1976). The industrial revolution in the home: Household technology and social change in the twentieth century. *Technology and Culture, 17,* 1–23. (Reprinted as "The Industrial Revolution in the Home," in D. MacKenzie & J. Wajcman, Eds., 1985, pp. 181–201.)

————. (1981). *More work for Mother: The ironies of household technology from the open hearth to the microwave.* New York: Basic Books.

Daly, Mary. (1978). *Gyn/Ecology: The metaethics of radical feminism.* Boston: Beacon Press.

————. (1984). *Pure lust: Elemental feminist philosophy.* Boston: Beacon Press.

Dawkins, Richard. (1976). *The selfish gene.* New York: Oxford University Press.

Dicken, P. (1992). *Global shift.* London: Paul Chapman.

Dinnerstein, Dorothy. (1977). *The mermaid and the minotaur: Sexual arrangements and human malaise.* New York: Harper Colophon Books.

Dworkin, Andrea. (1983). *Right-wing women.* New York: Coward-McCann.

Easlea, Brian. (1983). *Fathering the unthinkable: Masculinity, scientists and the nuclear arms race.* London: Pluto Press.

Eisenstein, Zillah. (1979). *Capitalist patriarchy and the case for socialist feminism.* New York: Monthly Review Press.

Elson, Diane. (1989). The cutting edge: Multinationals in the EEC textiles and clothing industry. In D. Elson & R. Pearson (Eds.), *Women's employment and multinationals in Europe* (p. 103). London: Macmillan.

Enloe, Cynthia. (1983). *Does khaki become you? The militarisation of women's lives.* London: Pluto Press.

————. (1989). *Bananas, beaches and bases.* Berkeley: University of California Press.

Fallows, J. (1981). *The national defense.* New York: Random House.

Fausto-Sterling, Anne. (1992). *Myths of gender.* New York: Basic Books.

Fee, Elizabeth. (1983). Women's nature and scientific objectivity. In Marian Lowe & Ruth Hubbard (Eds.), *Women's nature: Rationalizations of inequality* (pp. 9–27). New York: Pergamon Press.

Firestone, Shulamith. (1971). *The dialectic of sex.* New York: Bantam Books.

Frissen, Valerie. (1995). Gender is calling: Some reflections on past, present and future uses of the telephone. In K. Grint & R. Gill (Eds.), *The gender-technology relation: Contemporary theory and research* (pp. 67–92). London: Taylor and Francis.

Fuentes, A., & Ehrenreich, B. (1983). *Women in the global factory.* Boston: South End Press.

Gill, R., & Grint, K. (1995). Introduction. In K. Grint & R. Gill (Eds.), *The gender-technology relation: Contemporary theory and research* (pp.1–30). London: Taylor and Francis.

Green, E., Owen, J., & Pain, D. (1993). "City libraries": Human-centred opportunities for women? In E. Green, J. Owen, & D. Pain (Eds.), *Gendered by design: Information technology and office systems* (pp.11–30). London: Taylor and Francis.

Greenbaum, J. (1995). *Windows on the workplace: Computers, jobs and the organization of office work in the late twentieth century.* New York: Monthly Review Press.

Griffin, Susan. (1978). *Women and nature.* New York: Harper & Row.

Gunew, Sneja. (1990). *Feminist knowledge: Critique and construct.* New York: Routledge.

Gunnarsson, E., & Trojer, L. (Eds.) (1994). *Feminist voices on gender, technology, and ethics.* Lulea, Swed.: University of Technology Centre for Women's Studies.

Hacker, Sally. (1981). The culture of engineering: Woman, workplace, and machine. *Women's Studies International Quarterly, 4,* 341–353.

———. (1989). *Pleasure, power and technology.* Boston: Unwin Hyman.

Haraway, Donna (1978). Animal sociology and a natural economy of the body politic. *Signs, 4*(1), 21–60.

———. (1989). *Primate visions: Gender, race, and nature in the world of modern science.* New York: Routledge.

Harding, Sandra (1986). *The science question in feminism.* Ithaca, N.Y.: Cornell University Press.

———. (1993). Introduction. In Sandra Harding (Ed.), *The racial economy of science* (pp. 1–22). Bloomington: Indiana University Press.

———. (1998). *Is science multicultural? Postcolonialisms, feminisms, and epistemologies.* Bloomington: Indiana University Press.

Hartmann, Heidi. (1981). The unhappy marriage of Marxism and feminism. In Lydia Sargent (Ed.), *Women and revolution.* (pp. 1–41). Boston: South End Press.

Hawthorne, Susan, & Klein, Renate (Eds.). (1999). *Cyberfeminism: Connectivity, critique, and creativity.* Melbourne: Spinifex Press.

Healy, Bernadine. (1991, July 24/31). Women's health, public welfare. *Journal of the American Medical Association, 266,* 566–568.

Heeks, Robert B. (1993). Software contracting to the third world. In P. Quintas (Ed.), *Social dimensions of systems engineering: People, processes, policies and software development* (pp. 236–250). Hemel Hempstead, U.K.: Ellis Horwood.

Hrdy, Sarah. (1986). Empathy, polyandry, and the myth of the coy female. In R. Bleier (Ed.), *Feminist approaches to science.* (pp. 9–34). Elmsford, N.Y.: Pergamon Press.

Hubbard, Ruth. (1979). Introduction. In Ruth Hubbard & Marian Lowe (Eds.), *Genes and gender II: Pitfalls in research on sex and gender* (pp. 9–34). New York: Gordian Press.

———. (1990). *The politics of women's biology.* New Brunswick, N.J.: Rutgers University Press.

Jaggar, Alison. (1983). *Feminist politics and human nature.* Totowa, N.J.: Rowman & Allanheld.

Jaggar, Alison, & Rothenberg, Paula (Eds.). (1992). *Feminist frameworks.* New York: McGraw-Hill.

Keller, Evelyn F. (1982). Feminism and science. *Signs, 7*(3), 589–602.

———. (1983). *A feeling for the organism.* San Francisco: Freeman.

———. (1985). *Reflections on gender and science.* New Haven, Conn.: Yale University Press.

King, Ynestra. (1989). The ecology of feminism and the feminism of ecology. In J. Plant (Ed.), *Healing the wounds: The promise of ecofeminism* (pp. 18–28). Philadelphia: New Society.

Kirkup, G. (1992). The social construction of computers: Hammers or harpsichords? In G. Kirkup & L.S. Keller (Eds.), *Inventing women: Science, technology and gender* (p. 81). Cambridge, Eng.: Polity Press.

Knight-Ridder. (2000, April 11). Girls and computers, high-tech education. Philadelphia.

Kristeva, Julia. (1984). *The revolution in poetic language.* New York: Columbia University Press.

———. (1987). *Tales of love.* New York: Columbia University Press.

Maccoby, E., & Jacklin, C. (1974). *The psychology of sex differences.* Palo Alto, Calif.: Stanford University Press.

Macdonald, Anne L. (1992). *Feminine ingenuity: Women and invention in America.* New York: Ballantine Books.

MacKenzie, D., & Wajcman, J. (1985). *The social shaping of technology.* Milton Keynes, Eng.: Open University Press.

———. (1999). *The social shaping of technology.* (2nd ed.). Milton Keynes, Eng.: Open University Press.

MacKinnon, Catharine. (1982). Feminism, Marxism, and the state: An agenda for theory. *Signs, 7* (3), 515–544.

———. (1987). *Feminism unmodified: Discourses on life and law.* Cambridge, Mass.: Harvard University Press.

Merchant, Carolyn. (1979). *The death of nature.* New York: Harper & Row.

Microsyster. (1988). *Not over our heads: Women and computers in the office.* London: Microsyster.

Millar, Melanie S. (1998). *Cracking the gender code: Who rules the wired world?* Toronto: Second Story Press.

Mitter, S. (1986). *Common fate, common bond.* London: Pluto.

Mohanty, Chandra T. (1997). Women workers and capitalist scripts: Ideologies of domination, common interests, and the politics of solidarity. In M. Jacqui Alexander & Chandra T. Mohanty (Eds.), *Feminist genealogies, colonial legacies, democratic futures* (pp. 3–29). New York: Routledge.

National Science Foundation (1999). *Women, minorities, and persons with disabilities in science and engineering: 1998* (NSF 99-338). Arlington, Va.: NSF.

Norman, C. (1979, July 26). Global research: Who spends what? *New Scientist*, 279–281.

Office for Population Censuses and Surveys (OPCS). (1991). *Census of population.* London: HMSO.

O'Brien, Mary. (1981). *The politics of reproduction.* Boston: Routledge & Kegan Paul.

Paterson, Nancy. (1994). *Cyberfeminism.* Retrieved December 22, 2005, from http://internetfrauen.w4w.net/archiv/cyberfem.txt

Pattanaik, Bandana, (1997). *Feminist publishing in Asia.* http://www.spinifexpress.com.au/welcomeasia.htm.

Perrolle, Judith. (1987). *Computers and social change.* Belmont, Calif.:Wadsworth.

Plant, Sadie. (1996). On the matrix: Cyberfeminist simulations. In R. Shields (Ed.), *Cultures of the Internet: Virtual spaces, real histories, living bodies.* London: Sage.

Probert, B., & Hack, A. (n.d.) *Remote office work and regional development: The Australian Securities Commission in the La Trobe Valley.* Melbourne: CIRCIT.

Rich, Adrienne. (1976). *Of woman born: Motherhood as experience.* New York: Norton.

Rosser, Sue V. (1982). Androgyny and sociobiology. *International Journal of Women's Studies, 5*(5), 435–444.

———. (1988). Women in science and health care: A gender at risk. In Sue V. Rosser (Ed.), *Feminism within the science and health care professions: Overcoming resistance* (pp. 3–15). New York: Pergamon Press.

———. (1990). *Female friendly science.* New York: Pergamon Press.

———. (1993). Female friendly science: Including women in curricular content and pedagogy in science. *The Journal of General Education, 42*(3), 191–220.

———. (1994). *Women's health: Missing from U.S. medicine.* Bloomington: Indiana University Press.

———. (1998). The next millennium is here now: Women's studies perspectives on biotechnics and reproductive technologies. In Boel Berner (Ed.), *New perspectives in gender studies: Research in the fields of economics, culture and life sciences* (pp. 7–35). Stockholm, Swed.: Almquist and Wilosell International.

———. (2000). *Women, science, and society.* Elmsford, N.Y.: Pergamon Press.

Rothfield, Philipa. (1990). Feminism, subjectivity, and sexual difference. In Sneja Gunew (Ed.), *Feminist knowledge: Critique and construct* (pp. 121–144). New York: Routledge.

Sorenson, K. (1992). Towards a feminized technology? Gendered values in the construction of technology. *Social Studies of Science, 22*(1): 5–31.

Spanier, Bonnie. (1982, April). Toward a balanced curriculum: The study of women at Wheaton College. *Change, 14,* 31–34.

Stanley, Autumn. (1995). *Mothers and daughters of invention.* New Brunswick, N.J.: Rutgers University Press.

Suchman, L. (1994). Supporting articulation work: Aspects of a feminist practice of technology production. In A. Adam, J. Emms, E. Green, & J. Owen (Eds.), *Women, work and computerization: Breaking old boundaries—building new forms* (pp. 1–13). Amsterdam: North-Holland.

Suchman, L., & Jordan, B. (1989). Computerization and women's knowledge. In K. Tijdens, M. Jennings, I. Wagner, & M. Weggelaar (Eds.), *Women, work and computerization: Forming new alliances* (pp. 153–160). Amsterdam: North-Holland.

Tanesini, A. (1999). *An introduction to feminist epistemologies.* Oxford: Blackwell.

Thoresen, K. (1989). Systems development: Alternative design strategies. In K. Tijdens, M. Jennings, I. Wagner, & M. Weggelaar (Eds.), *Women, work and computerization: Forming new alliances* (pp. 123–130). Amsterdam: North-Holland.

Tong, Rosemarie. (1989). *Feminist thought: A comprehensive introduction.* Boulder, Colo.: Westview Press.

Trivers, R. L. (1972). Parental investment and sexual selection. In B. Campbell (Ed.), *Sexual selection and the descent of man.* Chicago: Aldine.

Wajcman, Judy. (1991). *Feminism confronts technology.* University Park: Pennsylvania State University Press.

Weber, Rachel. (1997). Manufacturing gender in commercial and military cockpit design. *Science, Technology and Human Values 22,* 235–253.

Webster, Juliet. (1995). *Shaping women's work: Gender, employment and information technology.* New York: Longman.

Wickham, J., & Murray, P. (1987). *Women in the Irish electronic industry.* Dublin: Employment Equality Agency.

Williams, Patrick, & Crissman, Laura. (1994). Colonial discourse and post-colonial theory: An introduction. In Patrick Williams & Laura Crissman (Eds.), *Colonial discourse and post-colonial theory* (pp. 1–20). New York: Columbia University Press.

Wilson, Edward O. (1975). *Sociobiology: The new synthesis.* Cambridge, Mass.: Harvard University Press.

Winner, Langdon. (1980). Do artifacts have politics? *Daedalus, 109,* 121–136.

Women Working Worldwide. (1991). *Common interests: Women organising in global electronics.* London: Women Working Worldwide.

Yerkes, Robert M. (1943). *Chimpanzees.* New Haven, Conn.: Yale University Press.

2

Women, Men, and Engineering

MARY FRANK FOX

Engineering is a revealing case in analyses of gender and status, and of challenges and prospects for women's occupational advancement. First, engineering is an important analytical case because it has been the most male-dominated profession in the United States. In 2001, women were 10.2 percent of all persons in the labor force with engineering degrees. They were 7.4 percent of those with doctoral degrees in engineering. In comparison, women were 43 percent of all college and university teachers, 29 percent of physicians, 23 percent of architects, and 20 percent of dentists employed in 2001 (CPST 2004, Tables 3.2 and 3.17).

Further, engineering is an important case for analysis because it is at the center of the development and applications of technology, and because it reveals a great deal about the relationship between technology and society. Engineering's connection with the development, operation, and maintenance of technological devices and processes has broad implications for industrial production, transportation, irrigation, housing, health care, data banks, energy, pollution, and environmental control (see Garrison, 1991; Grayson, 1977, 1993). In the twentieth century, engineering affected and reflected society through technological developments that include radar, navigation, microwaves, electronic instrumentation, ultrasound, and artificial organs (hearts, kidneys). In the twenty-first century, technology will continue to affect life

This chapter is an updated and revised version of "Women, Men, and Engineering," in *Gender Mosaics: Social Perspectives*, edited by Dana Vannoy, copyright © 2001. Reprinted with permission by Roxbury Publishing Company.

in the United States and throughout the world, particularly in the areas of energy, health care, and environmental controls. Technological innovations by themselves, however, do not have fixed or inevitable outcomes for society. Particular adoptions of technology, and the social relations that surround them, vary by time and place, and they depend upon complex factors of economy, politics, culture, and choice (for further discussion, see MacKenzie & Wajcman, 1999, pp. 1–16).

Nonetheless, because of the close relationship between engineering and technology, women's occupational status among doctoral-level engineers is significant to and telling of gender stratification in American society. Engineers with doctoral degrees are an important focal group for two reasons: first, because they deal directly with issues of technological research and its impact, and second, because they train university students, who, in turn, are critical for the future of science and society. In the analysis of gender stratification, women with doctoral degrees in engineering are significant because they have progressed through the proverbial educational pipeline (secondary, undergraduate, and graduate education) in science. They are a highly select group who have already overcome barriers of selection (both by themselves and by institutions) into engineering and have acquired credentials for high-level participation in the most male-dominated of professions. The questions of this chapter are: What is the profile of these doctoral-level women compared with men in participation, employment sector, fields, and rank—and what are the implications of this profile for gender, status, and engineering? What are the challenges and obstacles for women's attainments in engineering? Finally, what are the prospects and possibilities for the future?

Gender and Profile of Doctoral-Level Engineers

When one examines indicators of the career profiles of doctoral-level women and men in engineering, one can see that women and men differ in rates of participation, the places they work, the fields they occupy, and the ranks or positions they hold. Among doctoral-level women, further variation exists.

PARTICIPATION

Over the past half-century in the United States, women's and men's total participation in engineering is difficult to track precisely. Data on employment in science and engineering are not available systematically for this period. Data on doctoral degrees awarded are, however, and they indicate the proportions of women and men who have been professionally trained—and qualified at this level—for employment.

These data on gender and participation (Table 2.1) show the following patterns. First, in the decades of the 1950s and 1960s, women represented very low *proportions* of the total doctoral degrees awarded (less than half of 1 percent). During the 1960s, engineering and science fields experienced unprecedented increases in federal funding, degrees awarded, and positions created (Rossiter, 1995). But in that era of expansion, it was almost entirely men who were participating in engineering fields.

Second, in the decade of the 1970s—a decade significant for changes in gender composition in education and employment (Reskin & Roos, 1990)— women made gains in doctoral degrees awarded in engineering. Yet women still represented less than 2 percent of the total.

Third, during the 1980s and 1990s, changes occurred for women in engineering, a change somewhat later than that which occurred in other science fields (Fox, 1995). During the 1980s, women earned nearly 6 percent of all doctoral engineering degrees. Between 1990 and 1999, women earned 11 percent of them.

These data are for women across ethnic and racial groups. Relatively little is known about doctoral-level women in engineering by racial or ethnic status, in part because available data combine women (without differentiating by race or ethnicity) or combine racial and ethnic groups (without differentiating by gender) (Smith & Tang, 1994). What is known is that at the highest levels of education across science and engineering disciplines, non-Asian minorities are heavily underrepresented (Long & Fox, 1995, pp. 48–49; Smith & Tang, 1994, p. 121).

Table 2.1: PhD Degrees in Engineering, by Gender, 1950–99

Period	Men	Women	% Degrees Women
1950–59	5,731	13	0.2
1960–69	19,099	80	0.4
1970–79	29,990 ·	430	1.4
1980–89	30,416	1,914	5.9
1990–99	50,427	6,331	11.2

Source: *Professional Women and Minorities: A Total Human Resource Data Compendium,* 15th ed. (Washington, D.C.: Commission on Professionals in Science and Technology, 2004), calculated from table 2-160.

EMPLOYMENT SECTORS AND FIELDS

In sectors of employment, the vast majority of both men (87 percent) and women (84 percent) with doctoral degrees in engineering were clustered in academe and industry in 2001 (Table 2.2). However, locations differ somewhat by gender. Men are more likely than women to be employed in industry (Table 2.2).

Data available for fields within engineering are for those in which degrees are awarded, not for fields in which people are employed. The patterns of degrees earned by women compared with men vary by engineering field and period. One can look at the patterns in two ways: (1) by distributions of women's and of men's degrees by field and (2) by proportion of degrees awarded to women within fields.

Looking initially at the distribution of women and men holders of doctoral degrees by field (Table 2.3), one observes four patterns: (1) Across periods, women are less likely than men to be in mechanical engineering. (2) For both genders, the field with the greatest concentration of degrees was electrical engineering. However, from the 1980s to 1999, men were more highly concentrated in electrical engineering than were women. From 1990 to 1999, 28 percent of men, compared with 21 percent of women, were in electrical engineering. (3) As women earned increasing numbers of doctoral degrees in engineering in the 1980s and 1990s, they were more likely than men to be in chemical engineering. In the 1980s, 21 percent of women and 14 percent of men were in chemical engineering. From 1990 to 1999, 18 percent of women and 12 percent of men were in chemical engineering. (4) Across the decades,

Table 2.2: Employed Doctoral-Level Engineers, by Gender and Employment Sector, 2001

Sector	% Men (N = 91,980)	% Women (N = 7,610)
University/college	27.0	30.1
Other educational	0.6	1.4
Business and industry	59.8	54.3
Self-employed	2.6	2.0
Private-not-for-profit	2.7	3.5
Government	7.4	8.6
Other sectors	0.1	*
Total	100.0	100.0

*N suppressed—owing to fewer than 50 cases.
Source: *Professional Women and Minorities: A Total Human Resource Data Compendium,* 15th ed. (Washington, D.C.: Commission on Professionals in Science and Technology, 2004), table 3-20.

Table 2.3: PhDs Awarded in Engineering, by Fields and Gender, 1970–99

Field	1970–79			1980–89			1990–99		
	Men	Women	% Women	Men	Women	% Women	Men	Women	% Women
Aeronautical	1,444	16	1.1	1,162	39	3.2	2,228	124	5.3
% Total*	4.8	3.7		3.8	2.0		4.4	2.0	
Chemical	3,695	59	1.6	4,382	401	8.4	6,141	1,118	15.4
% Total	12.3	13.7		14.4	21.0		12.2	17.7	
Civil	3,696	47	1.3	3,984	219	5.2	5,575	698	11.1
% Total	12.3	10.9		13.1	11.4		11.1	11.0	
Electrical	7,139	102	1.4	7,129	309	4.2	14,294	1,353	8.6
% Total	23.8	23.7		23.4	16.1		28.3	21.4	
Industrial	938	21	2.2	883	111	11.2	1,855	351	15.9
% Total	3.1	4.9		2.9	5.8		3.7	5.5	
Mechanical	4,926	40	0.8	4,996	172	3.3	9,084	685	7.0
% Total	16.4	9.3		16.4	9.0		18.0	10.8	
Other	8,150	145	1.7	7,881	663	7.8	11,250	2,002	15.1
% Total	27.2	33.7		25.9	34.6		22.3	31.6	
Total	29,990	430		30,417	1,914		50,427	6,331	

* Percentages may not add to 100.0 due to rounding.

Source: Source: *Professional Women and Minorities: A Total Human Resource Data Compendium*, 15th ed. (Washington, D.C.: Commission on Professionals in Science and Technology, 2004), calculated from table 2-160.

women were more likely to be in industrial engineering, and by the 1990s, women and men were nearly equally likely to be in civil engineering.

Looking at data the other way—that is, by proportion of doctoral degrees awarded to women by fields (Table 2.3)—it is apparent that chemical and industrial engineering are the fields with the highest proportions of women. Aeronautical and mechanical engineering have the lowest proportions of doctoral degrees awarded to women. (The category "other fields" includes agricultural, bioengineering, ceramic, materials, mining and mineral, nuclear, and other fields of engineering outside those specified in the major categories in Table 2.3.)

These distributions of women and men by field suggest a nonuniformity in processes of selection, both the self-selection of women compared with men into various fields and the social restrictions that may operate with respect to gender on entry into and participation by field (Fox, 1995). Gender distributions by field are not unique to engineering; they operate in other disciplines as well. Much remains to be understood as to what governs and explains these distributions by field. Further, what is missing altogether from the data in Table 2.3 are the locations of women and men by *subfield*. Subfields are important because it is there that research is conducted. In contemporary science, traditional disciplines have split into hundreds and thousands of research subfields. These have come to blend areas that were formerly separate disciplines (e.g., biotechnology, bioengineering, and geophysics) (Grayson, 1993, p. 262; Hurd, 1997). Engineering research, as well, is strongly blended across fields and disciplines. The subfield of bioengineering, for example, has produced the pacemaker and defibrillator and has extended the life and health of heart patients during the past forty years.

It is important, therefore, to consider how research subfields in engineering are taking form, how they may be combining areas that were previously separate fields and disciplines, and what the implications of these patterns are. Significant concerns include the extent to which communication across fields influences engineering research; the ways in which the research advances with a mixture of theory, instrumentation, experimentation, and thinking in unconventional ways across areas; and the implications of trends in formation of subfields for the training and participation of women as well as men.

RANK

Except for the small group who are self-employed, people in engineering careers (and careers in other sciences) proceed through promotion in organizations. Across industrial settings, the variability of titles makes it difficult to determine patterns of gender and rank in industry. In academe, ranks are clearly, and in most cases uniformly, specified as professorial levels; so for the academic sector, one can assess patterns of rank by gender (Table 2.4).

The higher the professorial rank, the lower the proportion of women. As of 2001, across academic institutions within engineering fields, women comprised 15 percent of assistant professors, 9.4 percent of associate professors, and 2.8 percent of full professors (Table 2.4). In the distributions of women and men by their academic ranks (a second way of looking at the data in Table 2.4), we find inequalities particularly at the lower (assistant professor and below) and upper (full professor) ranks. That is, 51.6 percent of women (compared with 23.7 percent of men) were at ranks of assistant professor or lower (instructors and lecturers, including adjuncts and others); 17 percent of women (compared with 50.5 percent) of men were full professors. The proportions of women and of men at the associate level—31.3 percent of women and 25.7 percent of men—are more comparable.

It is important to consider what happens to the associate professors (the middle professorial level). In other disciplines—chemistry, mathematics, indeed across scientific disciplines, except for psychology—increase in the proportion of women full professors has not kept pace with the growth of women with doctoral degrees over time. Even allowing up to fifteen years from award of doctoral degree to rank of full professor, women's doctoral degrees have not translated into expected rank over time (Fox, 1999a). In

Table 2.4: Doctoral-Level Engineers in Academic Institutions, by Field and Gender, 2001

Rank	Men N	Women N	Percentage at each rank Men	Percentage at each rank Women
Professor	10,954	316	50.5	17.1
%	97.2	2.8		
Associate	5,563	577	25.7	31.3
%	90.6	9.4		
Assistant	4,170	730	19.3	39.6
%	85.1	14.9		
Instructor/ Lecturer	394	146	1.8	7.9
%	72.9	27.1		
Adjunct/Other	573	76	2.6	4.1
%	88.3	11.7		
Total	21,654	1,845	100.0	100.0

Source: *Professional Women and Minorities: A Total Human Resource Data Compendium,* 15th ed. (Washington, D.C.: Commission on Professionals in Science and Technology, 2004), calculated from table 3-246.

engineering, because the increasing numbers and proportions of doctoral degrees awarded to women are more recent, researchers will need to look at time spent as faculty in engineering, a subject to which this chapter will return in the section entitled "Prospects and Possibilities."

Challenges and Obstacles

In order to understand challenges for women's advancement in engineering, it is important to discern the relationship between gender and the roots and evolution of the field. The term *engineering* has its origins in the fifteenth century, when it first signified design of devices for warfare. In 1747, the first formal school of engineering, the Ecole Nationale des Ponts et Chases (National School of Bridges and Roads), was established in France. At the end of the eighteenth century, engineering began to acquire standing as a field of study, and the qualifier "civil" was applied to indicate nonmilitary focus and applications (see Ambrose et al., 1997, pp. 13–14). Nevertheless, engineering continued to be situated, directly and indirectly, within a military, and thus strongly masculine, model. In direct modeling, engineering education in the United States was founded in 1802 at the Military Academy at West Point. More indirectly, military institutions constructed technical "competition, merit, and especially discipline and control that defined both masculinity and success" (Hacker 1989, p. 66). In the nineteenth century, proficiency in mathematics came to be used as indicator—both cultural and technical—of rationality (Hodgkin, 1981) and self-discipline and "habits of industry," which were contrasted with passion, impulse, and instinct (Cardwell, 1957). More currently, engineering faculty who were interviewed about the importance of calculus and mathematics in engineering education believed that the subjects had "little relevance for job performance" but stressed that they were important "to show that you can do it" and "to develop a proper frame of mind" (Hacker, 1990, p. 149).

Images, symbols, and systems of belief have continued to link engineering with men and masculinity and separate it from women and femininity. Together, these symbols and systems operate to create a sense that such gender divisions are "natural" (Acker, 1999, p. 182)—with men as the standard group and women the nonstandard or "other" group who are "different" from the norm. Practices of promotion and reward, in turn, tend to advantage those who "look like" individuals currently in power. In engineering, as in other professions, these practices and conditions may favor women who have been able to construct a profile and personal circumstances that resemble those of men (see Bern, 1993) and who have been able to "assimilate" and "demonstrate loyalty to those who dominate" (Zweigenhaft & Domhoff, 1998). Despite the attainments of "exceptional women," however, gender divisions associated with engineering can support a claim that women are "different" from the standard, thus deterring the advancement of women as a group (Fox, 1999b).

Engineering, like other sciences, is strongly social and organizational work, and this presents challenges for the participation of underrepresented members, including women. Engineering, like other sciences, also involves the cooperation of people and groups; it requires human and material resources; in scope and complexity, it relies heavily on facilities, funds, apparatus, and teamwork (Fox, 1991, 1992). Scientific progress, more so than other forms of creativity, "relates to, builds upon, extends, and revises existing knowledge" (Garvey, 1979, p. 14). Productive scientific researchers must continually shape, test, and update their work, and this effort takes place interactively with others in the field (Fox, 1991). Active researchers gain information—and standing—relevant to their work in science through meetings, face-to-face interactions, sharing of preprints, and other means that allow them to generate ideas, engage interest, and evaluate work. If some groups are limited in access to social networks and exchange, this limitation restricts the possibilities for them not simply to participate in a social circle but, more fundamentally, to do research and show the marks of status and performance in the field (Fox, 1991). Data indicate that women, more so than men, are outside the circles of interaction and exchange central to the social processes of science and engineering. For example, compared with men, women in science are less likely to have professional connections as editors and journal referees, less likely to be asked to lecture or consult outside their institutions, and less likely to be invited speakers at professional meetings (see Fox, 1991, p. 195; Fox, 1996, p. 282). Furthermore, although women are as likely as men to coauthor their publications in science, women collaborate with fewer colleagues than do men (Fox, 1991, 1999). Tellingly, my national survey of more than 775 academic scientists in five fields, within electrical engineering and physics (the fields sampled with the lowest proportions of women), women reported a significantly lower frequency than men of "discussing research with departmental faculty" (in chemistry, computer science, and microbiology, the gender differences in frequency were not significant). Because face-to-face interaction activates research interests and reinforces work in science, the levels of interaction are consequential to professional participation, performance, and productivity.

Prospects and Possibilities

The changing worksites and skills of engineering have helped to shape women's increasing participation and prospects. Although engineering sites still include outdoor and industrial sites such as oil ranges, mines, bridges, roads, and other "field" locations, engineering work increasingly focuses on design, analysis, and computing conducted in office and laboratory settings that are relatively quiet, clean, and safe (see Garrison, 1991; Grayson, 1993; McIlwee & Robinson, 1992, p. 3). In addition, engineering has gained a prominent place in the management and direction of industry, with an increased emphasis on administrative and economic skills among engineers (Grayson, 1977). Design, analysis, and computing are not gender-neutral occupational emphases, but they are more androgynous and appealing to numbers of women (and many men) than engineering "fieldwork." Likewise, although the area of management is not gender neutral, it is increasingly attractive and rewarding for both women and men.

Among doctoral-level engineers, women's prospects as university faculty are potentially supported by the growing supply and very high quality of women engineering students and the importance of female faculty as teachers, advisers, and role models. Women faculty are consequential in science and engineering because, compared to men faculty, they act as primary research advisers for a larger number of female students, have more female students on their research teams, take a more structured (and equitable) approach in interaction with students, and put more emphasis upon giving help to advisees across technical as well as interactional capacities (Fox, 2003). Owing to strong competition between the industrial sector and academe to recruit doctoral-level engineering personnel (Fox & Stephan, 2001), as well as growing recognition that women faculty are important for recruitment and retention of high-quality students, the demand for women engineering faculty may increase. Demand could, in turn, have consequences for women's rewards, including rank as academic engineers.

Clearly, women in engineering are a highly selective and qualified group; their ability, educational credentials, and attainments are as high as or higher than those of their male counterparts. Thus, solutions of improved status for women are not a matter of correcting "personal deficits" (Fox 1996, pp. 283–285). Rather, because engineering is organizational work, subject to organizational signals, priorities, resources, and reward schemes, it is important to identify and attend to enabling or disabling features of the settings in which scientists and engineers study and work (Fox, 1991, 1996).

For example, my analysis of data from site visits to twenty-two science and engineering departments that had been more and less successful in proportions of degrees awarded to women over time revealed that departments consistently "high" or "improved" in degrees awarded to women had organizational characteristics that set them apart from departments that were consistently "low" in doctoral degrees awarded to women (Fox, 2000). Specifically, improved departments, especially, had a past or ongoing history of leadership on issues of participation and performance of women; high and improved departments had chairs and faculty who had given greater consideration to what constitutes a "good environment" for study; and departments in those fields with higher proportions of women were more likely to have written guidelines for evaluation of progress (Fox, 2000). Other data also show certain organizational features associated with equity, particularly clear and standardized criteria for evaluation, and open processes in hiring, promotion, and allocation of rewards (Evetts, 1996; Fox, 1991; Long & Fox, 1995). Additional factors that promote prospects for women in engineering and other sciences include placement of junior faculty in ongoing projects, and attention to range and scope of collaborative opportunities (Fox, 1991, 1996). The key point here is that just as organizations are structured for outcomes, so they can be restructured for greater equity and better use of talent of underrepresented groups.

Conclusion

Engineering is a critical field for the analysis of gender, status, and prospects for women because engineering has been the most male-dominated profession in the United States and because it is central to the development and applications of technology in society. Engineers with doctoral degrees are a significant focal group because issues of technological research and its impact, and the technological training of university students, are particularly pertinent for them.

The profiles of doctoral-level women and men engineers reveal notable gender differences in participation, sector of employment, field, and rank. They also show variation among women (and men). Although commonalties exist and persist in women's careers in engineering, for continuing research, it is also important to consider patterns of gender and status as they occur by employment sector, field, and subfields. The impact of social and organizational factors, such as patterns of interaction and workplace practices and policies, as they influence outcomes for women may vary by employment sector, field, subfield, and institutional locations in ways not yet apparent; so, too, may solutions for the advancement of women. The issues, challenges, and prospects of women's careers are complex considerations of where, in what settings, and under which conditions women do and do not attain significant participation and performance in engineering. And that is our venture for the future.

References

Acker, Joan. (1999). Gender and organizations. In Janet S. Chafetz (Ed.), *Handbook of the sociology of gender* (pp. 177–194). New York: Kluwer Academic/Plenum.

Ambrose, Susan A., Dunkle, Kristin L., Lazarus, Barbara B., Nair, Indira, & Harkus, Deborah A. (1997). *Journeys of women in science and engineering.* Philadelphia: Temple University Press.

Bern, Sandra Lipsitz. (1993). *The lenses of gender: Transforming the debate on sexual inequality.* New Haven, Conn.: Yale University Press.

Cardwell, Donald S. L. (1957). *The organization of science in England.* London: Heinemann.

Commission on Professionals in Science and Technology (CPST). (2004). *Professional women and minorities: A total human resource data compendium.* Washington, D.C.: CPST.

Evetts, Julia. (1996). *Gender and career in science and engineering.* London: Taylor and Francis.

Fox, Mary Frank. (1991). Gender, environmental milieu, and productivity in science. In Harriet Zuckerman, Jonthan Cole, & John Bruer (Eds.), *The outer circle: Women in the scientific community,* (pp. 108–204). New York: W. W. Norton.

———. (1992). Research productivity and the environmental context. In T. Whiston & R. Geiger (Eds.), *Higher education: The United Kingdom and the United States* (pp. 103–111). Buckingham, Eng.: Society for Research into Higher Education/Open University Press.

———. (1995). Women and scientific careers. In Sheila Jasanoff, Gerald Markle, James Petersen, & Trevor Pinch (Eds.), *Handbook of science and technology studies* (pp. 205–223). Thousand Oaks, Calif.: Sage.

———. (1996). Women, academia, and careers in science and engineering. In Cinda-Sue Davis, Carol Hollenshead, Barbara Lazarus, & Paula Rayman (Eds.) *The equity equation: Fostering the advancement of women in the sciences, mathematics, and engineering* (pp. 265–289). San Francisco: Jossey-Bass.

———. (1999a). Gender, hierarchy, and science. In Janet S. Chafetz (Ed.), *Handbook of the Sociology of Science* (pp. 441–457). New York: Kluwer Academic/Plenum Publishers.

———. (1999b, April 15). Gender, knowledge, and scientific styles. *Annals of the New York Academy of Sciences, 869,* 89–93.

————. (2000, Spring/Summer). Organizational environments and doctoral degrees awarded to women in science and engineering departments. *Women Studies Quarterly 28,* 47–61.

————. (2003). Gender, faculty, and doctoral education in science and engineering. In Lilli Hornig (Ed.), *Equal rights, unequal outcomes: Women in American research universities* (pp. 91–109). New York: Kluwer Academic/Plenum.

Fox, Mary Frank & Stephan, Paula. (2001). Careers of young scientists: Preferences, prospects, and realities by gender and field. *Social Studies of Science 31,* 109–122.

Garrison, Ervan. (1991). *A history of engineering and technology: Artful methods.* Boca Raton, Fla.: CRC.

Garvey, William. (1979). *Communication: The essence of science.* Oxford: Pergamon.

Grayson, Lawrence P. (1977). A brief history of engineering education in the United States. *Engineering Education 68,* 246–264.

————. (1993). *The making of an engineer: An illustrated history of engineering education in the United States and Canada.* New York: John Wiley.

Hacker, Sally L. (1983). Mathematization of engineering: Limits on women and the field. In Joan Rothschild (Ed.), *Machina ex dea: Feminist perspectives on technology* (pp. 38–58). New York: Pergamon.

————. (1989). *Pleasure, Power and Technology.* Boston: Unwin Hyman.

————. (1990). *"Doing it the hard way": Investigations of gender and technology.* Boston: Unwin Hyman.

Hodgkin, Luke. (1981). Mathematics and revolution from Lacroix to Cauchy. In Herbert Mehrtens, Henk Bos, & Ivo Schneider (Eds.), *Social History of Nineteenth Century Mathematics* (pp. 50–71). Boston: Birkhauser.

Hurd, Paul DeHart. (1997). *Inventing science education for the new millennium.* New York: Teachers College Press, Columbia University Press.

Long, J. Scott, & Fox, Mary Frank. (1995). Scientific careers: Universalism and particularism. *Annual Review of Sociology 21,* 45–71.

MacKenzie, Donald, & Wajcman, Judy. (1999). *The social shaping of technology* (2nd ed.). Buckingham, Eng.: Open University Press.

McIlwee, Judith S., & Robinson, J. Gregg. (1992). *Women in engineering.* Albany: State University of New York Press.

Reskin, Barbara, & Roos, Patricia. (1990). *Gender queues, job queues.* Philadelphia, Pa.: Temple University Press.

Rossiter, Margaret. (1995). *Women scientists in America: Before affirmative action, 1940–1972.* Baltimore, Md.: Johns Hopkins University Press.

Smith, Earl, & Tang, Joyce. (1994). Trends in science and engineering doctorate production, 1975–1990. In Willie Pearson, Jr., & Alan Fechter (Eds.), *Who will do science? Educating the next generation* (pp. 96–124). Baltimore, Md.: Johns Hopkins University Press.

Zweigenhaft, Richard L., & Domhoff, G. William. (1998). *Diversity in the power elite.* New Haven, Conn.: Yale University Press.

3

Still a Chilly Climate for Women Students in Technology: A Case Study

MARA H. WASBURN AND
SUSAN G. MILLER

The past two decades saw the implementation of a variety of programs that succeeded in attracting more women into the fields of science, engineering, and technology. Many of these women are now in highly visible positions. However, although women constitute 51 percent of the population of the United States and 46 percent of the labor force, only 23 percent of those who are employed in this country as scientists and engineers, across all degree levels, are women (Mervis, 2000; National Science Foundation, 2000). A July 2001 report released by the National Council for Research on Women finds that much of the progress women made in these areas has stalled or eroded. The report underscores the increasing need for a scientifically and technologically literate workforce as we enter the new millennium. One year earlier, the Morella Commission (Committee on Science, 2000), charged with developing strategies to attract more women and minorities into science, engineering, and technology, reported to the Committee on Science of the House of Representatives that significant barriers to attaining that goal are present from elementary school through college and beyond.

Women and girls will comprise at least half of the available science, engineering and technology talent pool. Therefore, it becomes imperative not only to attract but also to retain women and girls in these disciplines. Achieving this goal requires the work of formal programs established specifically to confront the individual, cultural, and structural factors that discourage women's access to science and closely related fields. These factors include definitions of gender-appropriate work that embody gender stratification in American society, an educational system that channels women away from these disciplines, and gender-based inequities in educational resources and opportunities in science and science-related disciplines (see Fox, 1999, pp. 441–447).

Male/female attitudes toward science and technology begin to differ as early as elementary and middle school and continue on into high school. It is during this period that girls develop an understanding of what social roles are appropriate for them (Seymour, 1999; Belenkey et al., 1986). They have reservations about the seemingly male "computer culture" as they watch boys utilizing computers for violent computer games and what they see as technology for its own sake (Margolis & Fisher, 2002; Welty & Puck, 2001; AAUW, 2000). There is little software that appeals to them. The tendency of boys to monopolize the computers is not being vigorously challenged (Borg, 1999). As a result, girls do not take advantage of after-school computer clubs or enroll in higher-level computer classes (Sanders, 1995).

Several factors work against girls' unequal participation in science, mathematics, and computer education. These include tracking, judgments about their ability, and limited access to qualified teachers and resources (Madigan, 1997; Weiss, 1994; Oakes, 1990). By the time they are at the point where they must choose careers, girls have less experience with computers and perceive that they are behind (Borg, 1999), decreasing their likelihood of entering the fields of science, engineering, and/or technology.

Today, there is a dearth of young women enrolled nationwide in secondary school computer science advanced placement classes. Their absence does not appear to stem from disinterest in computers but rather from applications that seem more attuned to the interests of boys (AAUW, 2000; Molad, 2000). Hence, young women entering colleges and universities in the areas of science, engineering, and technology are disadvantaged by their lack of computer experience (Sanders, 1995). They also appear to have career goals that are not as well defined as those of their male counterparts, and they often lack confidence in their abilities (Astin & Sax, 1996; Vetter, 1996). They encounter college and university classes that are unfriendly to them, impeding their learning. The absence of women faculty and mentors in the classroom and elsewhere, few women peers in their classes, and the lack of supportive networks can create a "chilly climate" for women in nontraditional fields. It is during this critical period that many women transfer into other fields (National Council for Research on Women, 2001; Seymour, 1999; Hanson, 1997; Seymour & Hewitt, 1997; Astin, 1993).

This chapter presents a case study of Women in Technology, a student organization at Purdue University, founded by the School of Technology administration in lieu of a funded, formal program comparable to Purdue's Women in Engineering, discussed below. The student organization was created to address a flat line in the growth of the number of women students over a five-year period. We describe the group and locate it within the broader analyses of programs, considering the range of feminist perspectives, meanings, definitions, and agendas/solutions driven by those definitions. To determine the extent to which the problems confronting Women in Technology are those identified in the research literature cited above, we examine the results of a survey of the members' attitudes, beliefs, and perceptions regarding their majors and intended careers, foregrounding the voices of the participants. We discuss focus groups in which the student members respond to the aggregate data from the survey, and then propose their own solutions to the problems identified in those data. Finally, we explore how a networking–mentoring and learning communities model for Women in Technology creates an environment for addressing their perceived problems as women pursuing entry into nontraditional careers.

A History of Women in Technology

Purdue University in West Lafayette, Indiana, is a Big Ten, Research I University with 37,871 students, 42 percent of whom are women, and 1,870 faculty members, 21 percent of whom are women. Purdue's School of Technology differs from the Schools of Engineering because of its hands-on, application-oriented programs. The Schools of Engineering programs require more rigorous mathematics and science instruction because a strong theoretical background is needed to develop and design new products. More important for present purposes, the Schools of Engineering have a formal Women in Engineering Program, which employs a full-time director and two assistant directors. The program offers a comprehensive array of educational enhancement activities for women who are interested in careers in engineering. Programming includes precollege outreach and recruitment activities, and a number of activities aimed at retaining both undergraduate and graduate women, including a formal mentoring program (Wadsworth, 2002). The Women in Engineering Program is credited with the fact that Purdue graduates more undergraduate women engineers than any other university in the nation.

Purdue's School of Technology consists of eight departments: Aviation Technology, Building Construction Management, Computer Graphics Technology, Computer Programming Technology, Electrical Engineering Technology, Industrial Technology, Mechanical Engineering Technology, and Organizational Leadership and Supervision. All are applied sciences.

The School of Technology enrolls 4,246 students, of whom 15.2 percent are women. Table 3.1 shows that during the past five years, the number of female faculty in the School of Technology has remained virtually unchanged, at only 12 percent of the total faculty.

Despite the growth of career opportunities for women in all areas of technology and concerted efforts to recruit women into the areas of science, engineering, and technology, Purdue University's School of Technology experienced no growth in the proportion of women students enrolled during the most recent five-year period, perhaps (in part) because it lacks a formal program comparable to the Women in Engineering Program described above. As shown in Table 3.2, women continue to represent only 15 percent of the school's student body.

Purdue University is comprised of ten schools. Table 3.3 shows that the School of Technology has the lowest proportion of women students, even lower than the Schools of Engineering. If the School of Technology's Department of Organizational Leadership and Supervision, which is a business-oriented program, is removed from the equation, the proportion of women drops to only 10 percent.

Table 3.1: Faculty by Rank and Gender at Purdue University, School of Technology

Rank	1997				2001			
	N	Men	Women	% Female	N	Men	Women	% Female
Prof.	23	21	2	8.7	27	25	2	8.0
Assoc.	59	53	6	10.2	59	52	7	11.9
Asst.	44	37	7	15.9	45	38	7	15.2
Total	126	111	15	11.9	131	115	16	12.2

Source: Office of Institutional Research, Purdue University

Table 3.2: Students by Gender at Purdue University, School of Technology

	1997		2001	
	Percentage	N	Percentage	N
Male	84.9	3,526	84.8	3,600
Female	15.1	629	15.2	646
Total	100.0	4,155	100.0	4,246

Source: Purdue University School of Technology Counseling Office

Table 3.3: Undergraduate Female Enrollments by School at Purdue University, Fall 2001

	% of Women
Veterinary Medicine	99
Education	82
Pharmacy, Nursing, & Health Sciences	76
Consumer and Family Sciences	69
Liberal Arts	62
Agriculture	46
Management	36
Science	35
Engineering	18
Technology	15
Technology (without the Department of Organizational Leadership and Supervision)	10

Source: Office of the Registrar, Purdue University

As noted earlier, the low numbers of women students in science, engineering, and technology classes, coupled with a scarcity of female role models and/or mentors, can create a chilly climate for the female students who remain in their programs (National Council for Research on Women, 2001; Seymour, 1999; Hanson, 1997; Seymour & Hewitt, 1997; Astin, 1993). Consistent with the literature about those disciplines, many of the women students in Purdue's School of Technology are feeling the effects of their isolation.

In an effort to address the small number of women students and to assist in their recruitment and retention, the School of Technology created Women in Technology as a student organization in December 1998. Its stated purpose was to promote the leadership of women students in the school through networking and mentoring. Seventy-five women from the School of Technology joined the new organization. Those applying for grants to support Women in Technology directed their efforts toward developing an assertiveness training program. The grant was funded, and the first assertiveness training program was held.

Feminist Theory

Rosser (1998) applied feminist theory to projects directed toward improving the position of women in science (and, by extension, in engineering and technology). She identified the most relevant features of nine feminist theories, arguing that different theories identify or emphasize different causes of the inequality of women and therefore suggest different programmatic approaches.

The activities of any given program do not necessarily reflect its theoretical perspective. For example, a program might adopt a *structural* definition of the problem yet engage in activities more in keeping with an *individual* definition as a perceived realistic adjustment to the strong resistance of the authorities it confronts to structural change (Fox, 1998). In other cases, possibly due to perceived ineffectiveness, those involved in a program might realize a disjunction between its theoretical perspective and its activities. Subsequently, they might initiate new activities better reflecting that perspective.

Using Rosser's level of analysis, Women in Technology at its inception could be characterized as operating at the intersection of liberal and existentialist feminism. By that we mean Women in Technology had the liberal feminist goal of promoting equal access by removing barriers (as is the case in most programs directed toward improving the position of women). Additionally, it appeared to have the existentialist feminist goal of enhancing the experiences of women students, as seen in the administrative decision to spend the group's first year dealing with what their leadership identified as a critical issue: assertiveness.

In 2001, we became faculty advisors to Women in Technology. We found an organization that seemed to be in disarray. The membership, which had initially stood at seventy-five women, had dropped significantly. Few women were attending meetings, and we were told that an assertiveness training programming scheduled for the spring semester had been cancelled. The first several meetings we attended seemed largely focused on process, with few interpersonal interactions occurring and little input from the members in attendance. Many of those members did not know one another. There was little agreement on what the organization's goals should be or on how to achieve the goals that had been set.

It was our belief that if we did not intervene, we would find ourselves presiding over Women in Technology's demise. As we saw it, the students were taking little responsibility for the organization, looking instead to faculty for direction. We could find no indication that the members had been consulted when setting the group's agenda. Research indicates that groups whose members actively participate in establishing group goals tend to be more committed to those goals than groups with "top-down" structures (Quinn, 1996; Schein, 1992). We were in agreement that if, in fact, Women in Technology were to survive, it would have to change from a faculty-driven to a student-driven organization.

One of the authors had been instrumental in creating networking–mentoring groups for women faculty and staff on the Purdue campus, as well as a research support group for doctoral students (Wasburn, 2001; Wenniger, 1997). Both groups contained elements of networking–mentoring and learning communities (see below). Building upon lessons learned from those groups, we believed that the networking–mentoring and learning communities model held promise for rebuilding Women in Technology.

A Networking–Mentoring and Learning Communities Model

Among the recommendations suggested by the National Council for Research on Women (2001) report was to support women undergraduates by investing in mentoring programs with role models who can put a human face on science, engineering, and technology. Two models for achieving these goals are networking–mentoring and learning communities. Organizing students into learning communities is a strategy that can connect students within a dauntingly large and lonely university campus (Shapiro & Levine, 1999). Learning communities can be organized around common interests and curricula. "These can be used to build a sense of group identity, cohesiveness, and uniqueness . . . and to counteract the isolation that many students feel" (Astin, 1985, p. 161).

Networking–mentoring has a long, rich tradition within academe as a strategy for bringing women together for their mutual benefit and support (Swoboda & Millar, 1986; Haring, 1999). Defined as "an ever-changing series of dyadic contacts in which each person plays the role of mentor or mentee to differing degrees in each dyad" (Swoboda & Millar, 1986, p. 11), networking–mentoring is an empowering strategy that has been successful in assisting women with academic progress as both faculty members and students. Its power comes from the fact that each woman involved in a networking–mentoring group spends some time as a mentor and some time being mentored, depending on the situation. Each member is encouraged to take leadership in areas where she has particular knowledge and/or interest and to seek mentoring in those areas where she is less knowledgeable, thus diminishing the hierarchy that can develop in organizations of this sort.

If this change were to be implemented, Women in Technology would now lie at the intersection of Rosser's (1998) framework of liberal, existentialist, and psychoanalytic feminism. It would still have the liberal feminist overarching goal of improving the position of women students. It would still have the existentialist feminist goal of enhancing their experiences, although the students would be the ones to determine exactly what those experiences should be. It would now incorporate features of mentoring and/or female role models, which are associated with psychoanalytic feminism. However, before we could be effective change agents for Women in Technology, we needed to gain an understanding of the women students who had chosen to join the organization.

METHOD

The following sections describe the study we conducted to help the School of Technology understand factors that contribute to the attitudes, beliefs, and perceptions regarding the majors and intended careers of the members currently involved with Women in Technology, and to help empower the organization to become self-directed.

Although the faculty and administration of the school are aware that the women comprise only 10 percent of the technology-related majors within the school, no formal study had been conducted to investigate student views or perceptions. In order to solicit student responses regarding these factors, a survey was administered to the current members of Women in Technology. Of the 81 students involved in the group, 51 responded to the survey, for a 63 percent response rate.

The survey questions were modified from the Women in Engineering Programs and Advocates Network (WEPAN) Pilot Climate Survey, designed to assess engineering students' perceptions of the educational climate at their universities (Brainard et al., 1993). Kramer (1996) identified those factors as isolation, the perceived irrelevance of theoretical preparatory courses, negative experiences in laboratory courses, classroom climate, and lack of role models. Santovec (1999) contends that the problem is one of image: that engineering and technology is not about helping society, a frequently cited desire of female students.

The survey consisted of both structured and open-ended questions. The nine structured questions, which appear in Table 3.4, were rated on a 5–point Likert scale, with responses ranging from strongly agree to strongly disagree. The first four questions focused on classroom climate, the next two questions on self-perception of technology abilities, and the last three questions on career choice.

In order to capture the students' lived experiences, we asked three open-ended questions dealing with their career choice satisfaction, gender-specific difficulties they may have encountered, and the support they felt they needed to be successful:

1. Are you pleased with your choice of a career in technology? Why or why not?
2. What, if any, gender-specific problems have you encountered as a woman in technology?
3. How can the School of Technology best support its women students?

These open-ended questions allowed for more specific and individualized responses and minimized the imposition of predetermined responses when gathering data (Gall, Borg, & Gall, 1996). It should be noted that both closed and open-ended questions reflect students' perceptions of the issues, which may or may not correspond to social reality as defined by others. In addition, we created five focus groups, each consisting of six Women in Technology members. The purpose was to ascertain students' reactions to the survey findings and hear what they thought might be done to address the problems the respondents to our survey had identified.

Patton (1990) recommends utilizing multiple methodologies when studying a phenomenon in order to strengthen the design. That process is termed "triangulation" (p. 187). To triangulate the data, we examine the Women in Technology documents: its constitution, program announcements, grant applications, and events brochures. However, we recognize that studying the members of Women in Technology limits the inferences we can draw from our data. Members of an organization self-select, which means they may be more committed to their careers. They may also have very different experiences from the rest of their cohort. Additionally, we have neither a control group nor comparable data from male technology students from which to draw comparisons.

Like all case study research, the objective of this study is to understand one particular situation; it is not research based upon sampling (Stake, 1995, p. 4). The inquiry was conducted in such a manner as to ensure that the subject was accurately identified and described and that limits on the transferability of the findings are specified (see Lincoln & Guba, 1985, p. 296; Marshall & Rossman, 1996, pp. 142–152). We believe the findings are consistent with the literature cited above, and that the power of the women's voices will permit some limited inferences to be drawn.

One final point: Because Women in Technology members come from such diverse departments as Organizational Leadership and Computer Programming Technology, we do not believe that findings are specific to any particular discipline, but that they can be applied to other nontraditional programs in which women are enrolled.

FINDINGS

Table 3.4 shows the Women in Technology Students' responses to the nine structured survey questions:

Consistent with the literature cited above, the responses indicate that a significant number of women still feel isolated in their classes. More than one-third of the women believe that the professors in their technology classes do not treat women and men equally. One-quarter of them do not feel comfortable going to their professors for assistance outside the classroom, and one-third are either uncertain or disagree with the statement that they feel like equal participants when working on group projects with male teammates.

At the same time, 70 percent of the women students feel confident in their technological abilities and that technology is an appropriate career for a woman. These findings indicate that individual-level definitions do not appear to be an issue among the student respondents. The problems confronting them appear to stem from their school's lack of a formal program effectively addressing problems that they share with their female classmates.

By far, the most compelling results of the survey are the voices that express the experiences of being a female student in the School of Technology. The responses expand upon the issues discussed above.

It is clear from the comments received that many female students feel outnumbered and even intimidated in class:

Table 3.4: Women in Technology Survey Responses to Structured Questions

Question	SA	A	U	D	SD
1. The professors in my technology classes treat women and men equally in the classroom.	20%	42%	6%	20%	12%
2. I am often one of only a few women in my technology classes.	33%	43%	4%	18%	2%
3. I participate equally in group projects with male teammates.	22%	44%	10%	22%	2%
4. I feel comfortable asking questions in class.	31%	37%	12%	18%	2%
5. I feel comfortable going to my technology professors for assistance outside the classroom.	20%	46%	10%	20%	4%
6. I feel confident in my abilities in my technology courses.	18%	52%	24%	6%	0%
7. I feel a technology career is an appropriate choice for women.	45%	25%	22%	8%	0%
8. My family supports my career choice.	55%	33%	4%	6%	2%
9. My friends support my career choice.	51%	31%	6%	12%	0%

n = 51

Nothing really specific. I just feel uncomfortable, like people are staring at me sometimes when I'm the only girl in the class. They probably aren't, but I feel like they are.

Being the only female in one of my lab classes is kind of nerve-racking.

The only problem I've had is being outnumbered like 10 to 1.

I'm the only female in a class of 20. You really feel alone.

I'm always one of the few females in class. That gives me a lot of pressure.

I had an experience in one of my classes where I felt that men thought the women in the class did not exactly belong there or were stupid.

My roommate freaked out one semester when the teacher made remarks about what she was doing in there with the guys. Something like, other girls would love to change places with her. What an idiot.

Particularly challenging for women students are the group projects that are a hallmark of science, engineering, and technology classes:

The men don't want us to work with them on projects. When we do, they give us these stupid jobs to do. If we say anything, they look at each other, so I just stopped saying anything.

My computer classes don't usually have any girls I know . . . when there are any other girls in there. I feel like all the male students know each other. I end up with the other girls whether or not I want to work with them.

When I have guys who don't want me in their groups or don't give me enough to do, I just shut up. If more women would do this, there wouldn't be so many problems. I think doing stuff like this just makes it worse for everyone.

I feel funny in some of the groups. The male students would rather be by themselves than with me. The professors don't help. They let everyone pick their own groups and don't think about what happens if you don't get picked. You have to walk up and say, "Can I work with you?"

Male students say things and do things that make them hard to deal with sometimes. They don't want you in their group.

I hate group projects. The guys don't really want to work with me. Most of the professors don't seem to care one way or the other. I think they'd all be happier if we just disappeared.

Some of the women surveyed appear to be demoralized by the male students in their classes who seem to believe men are more knowledgeable and express greater self-confidence. The women feel a lack of respect for their abilities by their male classmates:

> Computers are not enjoyable for me anymore because I feel stupid in my classes when guys overachieve in everything they do. They always answer all the questions.

> Men often times think that women are not knowledgeable with computers. They think they are more logical.

> Guys in my classes sometimes have that "I am better and can do it better" attitude. That really makes me mad.

> At times before they know you, men assume that you are less qualified or less intelligent than they are.

> I know some of the members think (or SAY they think) there's no problem. There is. There is a big problem. Ignoring it won't get rid of it.

It should be noted that some women students, though a decided minority, do react positively, seeing their small numbers as a challenge to succeed:

> It is a great feeling to excel in an area that is mostly males.

> It gives women a great opportunity to excel in a predominantly male career.

> Guys tend to take over the groups and try to do everything. It's fun though, because I know more than most of them. I just sit there and wait for them to mess up.

Many of the students believe that both the male and female faculty need education about issues concerning women students in mostly male classrooms:

> Some of my professors are really not so friendly to the girls in the class. I have a woman professor who definitely favors the guys. She always has something good to say about what they do, even when I know they didn't do as well as some of the rest of us.

> Some of the professors kind of roll their eyes when you ask a question. They don't seem to think you have any brains.

> Professors sometimes disrespect me in class. That hurts.

> I like some of the professors, but others just don't seem to care whether we learn or not. They act like this is something for guys and not for us.

The professors call on male students more than on us. I even had a women professor do that.

We need to help the professors to "get it" if you know at I mean. They need to treat us like they treat the guys, and I don't just mean the male professors. It's like they'd rather we weren't in the class.

Educate some of the faculty about female students. We're just as good, sometimes we're better.

They should not make girls feel stupid if they can't answer right away. Sometimes the women professors are just as bad. They get impatient. It takes me a while to think about answers. I don't want to get it wrong because I talked before I was ready.

Some of my computer classes are a little strange. I know what's going on, but the professor is so rushed that he doesn't even wait for me to answer sometimes. He goes and asks the person next to me, even when I tell him I know. It's like he thinks a boy will do it quicker. It's frustrating.

We need to learn to speak up for ourselves when we have professors who don't treat us equally. What should we say to them so that we don't get a bad grade? I feel funny saying anything when I know I may see the same professor next semester.

The students identify women mentors and role models as critical needs for women students in the School of Technology:

They need to get more women professors and encourage more women to be in technology.

Open the doors of the School of Technology and let more women faculty and students in.

I wish we had more women professors we could go to with our problems. We don't know what to do that won't make people angry with us. Who do you complain to?

I'd like to meet more women who have computer-type careers. Maybe then I'd get some idea of what I'd like to do.

I think that more women professors would be good. I don't think they give enough credit for the way they think and for their abilities.

I think I would have liked to have a mentor, maybe someone from business. I didn't meet any female professors, especially in my computer classes, who seemed to care. I'm changing majors. I think I might have stayed if I had someone to talk to.

Don't worry so much about getting people to *come* to the school. Worry about what happens to us when we *get* here. I hope there's going to be a mentoring program. That's what you should really do. Get mentors for the students that are here already.

To triangulate our data, we reviewed the Women in Technology constitution, the preamble of which states that Women in Technology is dedicated to "promoting the leadership of women through networking, encouragement, mentoring, and outreach." However, those applying for grants to support the organization directed their efforts toward developing assertiveness training programs. These were to emphasize "leadership and assertiveness, conflict management, and negotiation skills" (AlliedSignal grant application, March 26, 1999).

We also reviewed the membership lists (which showed a steady decline), program announcements, and grant applications—in particular the grant to support the Women in Technology Assertiveness Training program described above. The grant applications stated that "women and men in team projects need to be more assertive. Women tend to think their behavior is already assertive while their peers would disagree and label women's behavior and communication skills as unassertive. Male students generally reported that they knew they were being unassertive" (AlliedSignal Grant Application, March 16, 1999). As described in the grant application, the program would consist of workshops on conflict management, sexual harassment, and interpersonal communication.

The assertiveness training on which Women in Technology was to focus its first year reflects a theoretical perspective, shaping the way in which the problem was defined and a solution sought (Fox, 1998). The group was using an individual definition, in which "the status of women is attributed to, or said to correspond to, women's individual characteristics, such as attitudes, behaviors, aptitudes, skills, performance, and experience" (Fox, 1998, p. 202). However, as noted above, when the students were given the opportunity to assess their own problems and develop strategies to address them, they reported many structural causes for the inequities they continue to experience, creating a combined environmental/individual definition. Hence, looking at the organization's stated purpose and the operationalization of that purpose, there appears to be a disjunction between the two. The creation of an assertiveness training program was responsive neither to the organization's stated goals (noted above) of "networking, encouragement, mentoring, and outreach" nor to the students' concerns emerging from the study.

There is, perhaps, a caveat with regard to the small numbers of women students and faculty in the School of Technology. Rosser (1997) observes that the overwhelmingly male nature of science has not only afforded women limited access to participation in the discipline, but it has shaped the very nature of the discipline itself. She notes that science is neither unbiased nor value-free. Everything from what is studied to the subjects for the experiments is male-dominated. When women begin to enter a field, they ask new questions and challenge existing assumptions (Rosser, 1997, p. 84). Similarly, the entry of large numbers of women into the School of Technology may change the way in which technology is studied and developed—perhaps the very nature of technology itself.

FOCUS GROUPS

At the final Women in Technology meeting of the semester, we shared the results of the survey with the students, taking time to respond to their questions. We also discussed some of the support structures available in the Women in Engineering program, many of which are directed at ameliorating the same problems they face as women technology students.

Although the members were keenly aware that women were a minority within the School of Technology, many of them were shocked by exactly how few women faculty and students there are. They were also relieved to hear how other members of the group shared their struggles. They spent some time sharing their experiences and questioning one another about various classes they had taken. For the first time, it appeared to us that the members were connecting with one another on a personal level. Lacking a formal program to support them, the solutions to their problems would require collective effort.

The literature points to the sense of isolation that women technology students experience as a result of their small numbers. A self-directed model for Women in Technology would provide a vehicle for the group to fulfill its mission of recruiting women into the school and creating a cohesive organization to help retain the women who enter.

When the conversations subsided, we discussed some of the research surrounding empowerment through student-driven, rather than faculty-driven, organizations, and some of the theory behind networking–mentoring and learning communities. Then, we divided the students into focus groups and asked them to identify strategies that would address the concerns revealed in the surveys. Seven suggestions for the group emerged from those discussions:

1. Invite the women professors to dinner so that everyone can get to know each other.
2. Create a living–learning center so that women students in technology can study together, live together, and take classes together. (There are other such centers on campus.)
3. Begin a mentoring program for all students. This could include women mentors already in the workplace, especially ones who have gone through this program.
4. Create support groups within departments in order to discuss common problems and concerns and how to deal with them.
5. Have a retreat so that the members will really get to know one another. Keep networks between members open.
6. Create an outreach program with area high schools to encourage more women to go into technology careers.
7. Devote one Women in Technology meeting a month to discussing the issues from the survey, such as how to get more women students into the School of Technology, and how to make women more comfortable in the school.

In addition, there was one suggestion for us as their faculty advisors: educate the faculty about issues affecting women students in the school. Historically, this strategy has been problematic as a vehicle for change. In the 1920s, despite numerous surveys and copious statistical evidence, women faculty quickly discovered that they had no power to force universities to hire women; their efforts at persuasion failed (Rossiter, 1995, p. 161). In the 1940s, the American Association of University Women passed a series of resolutions condemning quotas for women in undergraduate and graduate education, as well as in employment. However, they had no power to enforce those resolutions. Again, persuasion was their only tool, and again, there were no improvements (Rossiter, 1995, p. 40). It was not until the 1960s and 1970s, when women like sociologists Sylvia Fleis Fava and Alice Rossi began to publish research articles documenting the plight of academic women, that changes began, ultimately leading to the passage of Title IX (Rossiter, 1995, pp. 161–182). Yet inequities persist.

On the Purdue campus, the response to data on gender-based inequities has been somewhat more positive. In 1988, a task force of faculty women (under the auspices of the Faculty Affairs Committee of the University Senate) conducted a needs assessment to determine the support for an expanded Women's Resource Office on campus. They invited all women faculty to submit a letter outlining concerns relevant to women at Purdue. In their responses, words such as *chilly . . . nonsupportive . . . unsympathetic . . . hostile . . . isolating . . . deplorable . . . disrespectful . . . sexually harassing . . .* were used to describe the campus climate (Jagacinski, 1988, p. 1).

Subsequent to the publication of a needs assessment report, the university's upper administration created and implemented a number of policies to address the problems that the assessment had identified. Most important, as the direct result of the presentation of the data, the Women's Resource Office, which had been at the center of the controversy, was fully funded, and the position of a full-time director was created.

Comparable data have influenced administrators at other universities as well. For example, in his introductory remarks to a formal report on the status of women on the science faculty at the Massachusetts Institute of Technology, MIT President Charles M. Vest noted, "I have always believed that contemporary gender discrimination within universities is part reality and part perception. True, but I now understand that reality is by far the grater part of the balance" (Massachusetts Institute of Technology, 1998, p. 1). He subsequently initiated policies to attract and retain women faculty and students in the School of Science at MIT.

Given this precedent, we shared these data with many of our male colleagues. They agreed that educating male students and faculty members will be a critical part of the effort to retain women students. It is our hope that through various efforts, including a series of workshops for male and female faculty members and graduate student instructors entitled "Making Your Classroom Woman-Friendly," we can continue to find more ways to make women students welcome in our classrooms.

IMPLICATIONS

We will present these student recommendations to the membership as a whole and support their efforts to implement them, measuring the group's effectiveness by how much leadership the members show in directing the organization—specifically, the number of programs they present that address the concerns of the group, and the number of active members who attend meetings. As is the case with all student-run organizations, the turnover of the membership each year threatens the continuity of initiatives, potentially limiting the group's effectiveness.

An ongoing effort that seeks to understand the student issues and address their needs is a necessary first step, followed up by an assessment of the progress that has been made toward greater attraction and retention of women students in the School of Technology, and a plan to target areas that still need to be improved. Above all, we will continue to listen to the voices of our women students and to involve them in all of our efforts. Their voices express confidence in their own abilities in technology courses, belief that technology is an appropriate career choice for women, and a feeling that their families and friends support their career choices. At the same time, the problems they share have persisted and require structural solutions—that is, the development of enduring programs.

It is important to remember that it is the students who are directly affected by the problems they identify, so they must be involved in any solution. In the final analysis, however, these problems are tied to challenges we face as a nation. The more skills that are developed without regard to gender, the more technologically equipped members there will be in our society, and the more productive our society will be. In the present world economy, this is far from a trivial matter.

References

American Association of University Women. (2000). *Tech-savvy: Educating girls in the new computer age.* Washington, D.C.: AAUW Educational Foundation.

Astin, A. W. (1985). *Achieving educational excellence.* San Francisco: Jossey-Bass.

———. (1993). *What matters in college? "Four critical years" revisited.* San Francisco: Jossey-Bass.

Astin, H. S., & Sax, L. J. (1996). Developing scientific talent in undergraduate women. In C. S. Davis, et al. (Eds.), *The equity equation: Fostering the advancement of women in the sciences, mathematics, and engineering* (pp. 96–121). San Francisco: Jossey-Bass.

Belenkey, M. F., et al. (1986). *Women's ways of knowing.* New York: Basic Books.

Borg, A. (1999). What draws women to and keeps women in computing? In C. C. Selby (Ed.), *Women in science and engineering: Choices for success* (pp. 102–105). New York: New York Academy of Sciences.

Brainard, S. G., et al. (1993). WEPAN pilot climate survey. Retrieved September 16, 2001, from http://www.wepan.org (the Women in Engineering Programs and Advocates Network Web site).

Committee on Science, House of Representatives. (2000). *A review of the Morella Commission report recommendations to attract more women and minorities into science, engineering, and technology* (Serial No. 106–83). Washington D.C.: U.S. Government Printing Office.

Fox, M. F. (1998). Women in science and engineering: Theory, practice, and policy in programs. *Signs: Journal of women, culture, and society, 24*(1), 201–223.

———. (1999). Gender, hierarchy, and science. In J. S. Chafetz (Ed.), *Handbook of the sociology of gender* (pp. 441–458). New York: Klewer/Plenum Publishers.

Gall, M. D., Borg, W. R., & Gall, J. P. (1996). *Educational research: An introduction* (6th ed.). New York: Longman.

Hanson, S. L. (1997). *Lost talent: Women in the sciences.* Philadelphia: Temple University Press.

Haring, M. J. (1999). The case for a conceptual base for minority mentoring programs. *Peabody Journal of Education, 74*(2), 5–14.

Jagacinski, C. (1988). *Report on the status of women faculty.* West Lafayette, Ind.: Purdue University, Women's Resource Office.

Kramer, P. (1996). Engineering up front: Why hands-on engineering education works for women and girls. *GATES, 3*(1), 39–44.

Lincoln, Y., & Guba, E. (1985). *Naturalistic inquiry.* Beverly Hills, Calif.: Sage Publications.

Madigan, T. (1997). *Science proficiency and course taking in high school: The relationship of science course-taking patterns to increases in science proficiency between eighth and twelfth grades* (NCES 97–838). Washington, D.C.: U.S. Department of Education, National Center for Education Statistics.

Margolis, J. & Fisher, A. (2002). *Unlocking the clubhouse: Women in computing.* Cambridge, Mass.: MIT Press.

Marshall, C., & Rossman, G. B. (1995). *Designing qualitative research* (2nd Ed.). Thousand Oaks, Calif.: Sage Publications.

Massachusetts Institute of Technology. (1998). Women faculty in science at MIT [Special edition]. *The MIT Faculty Newsletter, XI*(4), 1–13.

Mervis, J. (2000). Diversity: Easier said than done. *Science, 289*(5478), 378–379.

Molad, C. B. (2000). *Women weaving webs: Will women rule the internet?* Houston: CBM Press.

National Council for Research on Women. (2001). *Balancing the equation: Where are women and girls in science, engineering and technology?* New York: National Council for Research on Women.

National Science Foundation. (2000). *Science and engineering indicators* (Appendix Table 3–10: A-155). Washington, D.C.: National Science Foundation.

Oakes, J. (1990). *Multiplying inequalities: The effects of race, social class, and tracking on opportunities to learn mathematics and science.* Santa Monica, Calif.: The Rand Corporation.

Patton, M. Q. (1990). *Qualitative evaluation and research methods* (2nd ed.). Newbury Park, Calif.: Sage Publications

Quinn, R. E. (1996). *Deep change: Discovering the leader within.* San Francisco: Jossey-Bass.

Rosser, S. V. (1997). *Re-engineering female friendly science.* New York: Teachers College Press.

———. (1998). Applying feminist theories to women in science programs. *Signs: Journal of women, culture, and society, 24*(1), 171–199.

Rossiter, M. W. (1995). *Women scientists in America: Before Affirmative Action, 1940–1972.* Baltimore: The Johns Hopkins University Press.

Sanders, J. (1995). Girls and technology: Villain wanted. In S. V. Rosser (Ed.), *Teaching the majority: Breaking the gender barrier in science, mathematics, and engineering* (pp. 147–159). New York: Teachers College Press.

Santovec, M. (1999). Campus climate affects female engineering undergrads. *Women in Higher Education, 8*(7), 5.

Seymour, E. (1999). The role of socialization in shaping the career-related choices of undergraduate women in science, mathematics, and engineering majors. In C. C. Selby (Ed.), *Women in science and engineering: Choices for success* (pp. 118–126). New York: New York Academy of Sciences.

Seymour, E., & Hewitt, N. H. (1997). *Talking about leaving: Why undergraduates leave the sciences.* New York: Westview Press.

Schein, E. H. (1992). *Organizational culture and leadership* (2nd. ed). San Francisco: Jossey-Bass.

Shapiro, N. S., & Levine, J. H. (1999). *Creating learning communities.* San Francisco: Jossey-Bass.

Stake, R. E. (1995). *The art of case study research.* Thousand Oaks, Calif.: Sage Publications.

Swoboda, M. J., & Millar, S. B. (1986). Networking–Mentoring: Career strategy of women in academic administration. *Journal of NAWAC, 8–13.*

Vetter, B. M. (1996). Myths and realities of women's progress in the sciences, mathematics, and engineering. In C. Davis, et al. (Eds.), *The equity equation: Fostering the advancement of women in the sciences, mathematics, and engineering* (pp. 29–56). San Francisco: Jossey-Bass.

Wadsworth, E. M. (2002). *Giving much, gaining more: Mentoring for success.* West Lafayette, Ind.: Purdue University Press.

Wasburn, M. H. (2001). Creating a community to help ABDs graduate. *Women in Higher Education, 10*(1), 36–37.

Weiss, I. R. (1994). *A profile of science and mathematics education in the United States: 1993.* Chapel Hill, N.C.: Horizon Research.

Welty, K., & Puck, B. (2001). *Modeling Athena: Preparing young women for citizenship and work in a technological society.* University of Wisconsin–Stout.

Wenniger, M. D. (1997). New networking model of mentoring catches on at Purdue. *Women in Higher Education, 6*(10), 1–2.

4

The Feminization of Work in the Information Age

JUDY WAJCMAN

The future of work and the transformation of family life are key issues in contemporary social science. Many believe that the invention and diffusion of digital technologies are factors at the heart of these transformations. Much emphasis is placed on major new clusters of scientific and technological innovations, particularly the widespread use of information and communication technologies (ICTs), and the convergence of ways of life around the globe. The increased automation of production and the intensified use of the computer are said to be revolutionizing the economy and the character of employment. In the "information society" or "knowledge economy," the dominant form of work becomes information and knowledge-based. At the same time, leisure, education, family relationships, and personal identities are seen as molded by the pressures exerted and opportunities arising from the new technical forces.

Theorists of the information society have, for the most part, shown little interest in the question of changing gender relations. This, despite the fact that the feminization of the labor force has been heralded as one of the most important social changes of the twentieth century. Their oversight is particularly disappointing given the contribution that feminist scholarship has made to our understanding of the gendered nature of both work and technology. The sexual division of labor is central to technological development and the organization of work. Indeed, the literature demonstrates that the relationship between gender relations and ICTs is a complex mixture of interactive processes, a key site of which is the workplace (Wajcman, 2004).

The challenge, then, is to untangle the relationship between these dynamic processes and assess the implications for gender equality. The key underlying question is: to what extent are the older hierarchies of the gender order being destabilized in the digital economy? We begin by considering the meaning of the term "information society" and examine what the shift to a service economy means for women's position and experiences in the labor market. The chapter then looks at the implications of organizational restructuring for women's careers and power relations in management. Finally, we will look at how ICTs affect the spatial organization of economic activity.

The Information Society

The notion of the information society has been absorbed into everyday use, and it is also widely criticized (Castells, 1996; Loader, 1998; Webster, 1995). I use the term here because it captures the evolutionary and determinist frameworks employed by most theorists of the "information age." These frameworks are not only gender blind, but they also fail to recognize the operation of wider power relations. Many social commentators have written with confidence about the technological, economic, social, and cultural changes over the last two decades (Kumar, 1995; Held, Goldblatt, & Perraton, 1999). Several different schools of thought exist about postindustrial society, but the recurring theme is a claim that theoretical knowledge and information have taken on a qualitatively new role. Prominence is given to the intensity, extent, and velocity of global flows, interactions, and networks embracing all social domains.

One of the best known commentators of such change is Manuel Castells (1996), who argues that the revolution in information technology is creating a global economy, the product of an interaction between the rise in information networks and the process of capitalist restructuring. In the "informational mode of development," labor and capital, the central variables of the industrial society, are replaced by information and knowledge. In the resulting "Network Society," the compression of space and time made possible by the new communication technology alters the speed and scope of decisions. Organizations can decentralize and disperse, with high-level decision making remaining in "world cities" while lower level operations, linked to the center by communication networks, can take place virtually anywhere. "Information" is the key ingredient of social organization, and flows of messages and images between networks constitute the basic thread of social structure (Castells, 1996, p. 477). For Castells, the information age, organized around "the space of flows and timeless time," marks a whole new epoch in the human experience.

While optimistic and pessimistic visions of the information age exist, they all focus on the assumed outcomes for employment. The optimists see the expansion of the information-intensive service sector as producing a society based on lifelong learning and a knowledge economy. This implies that a central characteristic of work will be the use of expertise, knowledge, judgment, and discretion in the course of producing a product or service, requiring employees with high levels of skills and knowledge. The pessimistic neo-Braverman approach, by contrast, sees growing technology-induced unemployment and increased vulnerability to global capital (see review by Burris, 1998). It argues that automation standardizes worktasks and diminishes the need to exercise analytical skills and theoretical knowledge.

What is common to most of these understandings of the new social order is their tendency to adopt a technologically determinist stance. Castells explicitly builds on theories of postindustrialism, moving beyond a teleological model and giving the analysis a global reach (Bell, 1973). However, while he explicitly attempts to distance himself from technological determinism, he does not entirely succeed. The idea that technology, specifically information and communication technology, is the most important cause of social change permeates his analysis of Network Society. Similarly, there is a tendency to conceptualize these technologies in terms of technical properties and to construct the relation to the social world as one of implications and impacts. The result is a rather simplistic view of the role of technology in society. In this, Castells is typical of most scholars of the information society who fail to engage with the burgeoning literature in the social studies of science and technology (STS) that has developed over the last two decades (Jasanoff et al., 1995; MacKenzie & Wajcman, 1999).

While space is not available here to engage in any depth with this literature, the point of the STS literature is not to deny the transformative potential of technology. Rather, STS emphasizes that technological change is itself shaped by the social circumstances within which it takes place. STS studies show that the generation and implementation of new technologies involve many choices between technical options. A range of social factors affects which of the technical options are selected. These choices shape technologies and, thereby, their social implications. In this way, technology is a socio-technical product, patterned by the conditions of its creation and use. Understanding the place of these new technologies from such a perspective requires avoiding a purely technological interpretation and recognizing the embeddedness and the variable outcomes of these technologies for different social groups. Technology and society are bound together inextricably; this means that power, contestation, inequality, and hierarchy inscribe new technologies. ICTs can indeed be constitutive of new social dynamics, but they can also be derivative or merely reproduce older conditions. Moreover, it is increasingly recognized that the same technologies can have contradictory effects.

From Manufacturing to Services

Given the complexity of the relationship between technology and social change, it is apparent that the effects of ICTs specifically on women's work are extremely hard to isolate and assess. Many of the employment dynamics discussed in this chapter are functions of restructuring and work reorganization within industries. As Webster (2000, p. 123) notes, "The introduction and application of ICTs are a part of, as much as the consequences of, technological change." It is clear, however, that ICTs are profoundly implicated in changes in work reorganization and work location in particular industries that are central to the information society. What, then, are some of the key changes to women's work in the information society?

One of the most striking features of advanced capitalist economies is the feminization of the labor force. The sharp separation between work and home has been eroded by this development, at least for some social strata. The most important change to women's access to paid employment has been the sharp increase in the labor market participation of married women and women with young children. The causes of feminization are complex, but clearly they link to the substantial growth in service-sector activity and employment. In most advanced countries, the manufacturing sector has declined, with most new jobs being created in services. In one sense, this advantages women, since they have long been associated with service work, especially jobs involving caring for and catering to the needs of clients. Women have predominated numerically in clerical work, retail, catering, and the health and education professions, all of which are important providers of jobs in the modern service-based economy (Bradley et al., 2000; Webster, 1999).

Accompanying the feminization of the labor force has been a dramatic growth in economic inequality between different groups of women. This phenomenon is especially striking in the United States, where feminist demands for gender equality have been more potent than elsewhere, and where inequality in wealth and income has increased sharply in recent years (Jacobs, 1995).

Although this trend to feminization has not brought about a major breakdown of gender segregation, there has been a significant movement of women into traditionally male professions such as law, medicine, and management. This elite of women has unprecedented access to well-paid, high-status occupations, while at the bottom of the occupational hierarchy, women have expanded their share of already femininized lower-skilled or lower-paid occupations. Indeed, in many traditionally female "semi-professions" like nursing and secretarial work, restructuring is creating a small minority of highly paid, highly skilled workers alongside a much larger number of poorly paid, minimally trained support workers. In these latter cases, polarization is occurring within a virtually all-female occupational category (Milkman, 1995). At the other end of the spectrum, advanced capitalist countries have seen an enormous growth among "contingent" workers, the majority of whom are women. Women make up the majority of part-time workers, temporary workers, and at-home "independent contractors" (Rubery et al., 1998).

The increase in "flexible" or "atypical" work that characterizes this era of economic globalization could not have occurred without the proliferation of ICTs that support it. For example, the financial service and telecommunications industries rely heavily on information technology (IT) for service and sales delivery. Business transactions ranging from personal retail banking to transnational financial market deals are increasingly mediated by IT. Researchers generally note that IT has three distinct features that can change dramatically the way service work is organized: the "automate, informate and networking capabilities" (Yeuk-mui, 2001, p. 178). Automation means the replacement of manual labor by the IT system to accomplish menial worktasks. The informating ability refers to the capability of IT to generate detailed information about the work process. The networking ability of IT refers to the use of company-wide intranets to coordinate worktasks, disseminate information, and exchange opinions between employees in different ranks and functional departments. These three distinct capabilities of IT have been widely acknowledged (see Zuboff, 1988). However, little consensus exists regarding their impact on work organization and how they affect the experience of women in employment.

One issue that has occupied feminist scholars has been the growing concern in contemporary firms with quality and customer service. Because of the complex interplay of technological and organizational developments, service work increasingly relies on "front-line" work that is people oriented. It has become common to argue that service work challenges our usual conceptions of work because the quality of the service is so intimately related to the personal qualities and social skills of the service providers (Leidner, 1993; Du Gay, 1996). Service work is being feminized in more than simply numerical terms, in that to an increasing extent these jobs require the supposedly feminine qualities of serving and caring. These new forms of labor have been variously theorized as "emotional labor" (Hochschild, 1983), "sexual labor" (Adkins, 1995), and "aesthetic labor" or "body work" (Tyler & Abbott, 1998).

The feminization of service work has specific implications for women because their physical appearance and "personality" become an implicit part of the employment contract. An illustration is the requirement imposed on female, but not male, flight attendants to weigh in periodically during routine grooming checks in order to maintain a strict weight–height ratio (Tyler & Abbott, 1998). The enforcement of weight standards by the airline industry leads to "enforced" dieting in pursuit of the thin, ideal body. This aesthetic labor or bodywork, like emotional and sexual labor, is an integral part of the "effort bargain" (Forrest, 1993). Yet it is not recognized or remunerated because it is seen as what women *are* rather than what women *do.*

Overall then, the shift from manufacturing to services must be viewed as a mixed blessing for women. On the one hand, women are more fully integrated into the paid labor force and are unlikely to be relegated to the domestic sphere of yesteryear. On the other hand, many of the jobs created in this sector are temporary and part-time. While these jobs offer women more flexibility, the use of IT by employers to fine tune their labor requirements can cost women dearly in terms of pay, conditions, and training opportunities. The skill requirements for much service sector employment tend to be social and contextual, making them less amenable to formal measurement. The issue of how skills or competencies are perceived, labeled, accredited, and rewarded is critical for women's ability to participate in and benefit from the "knowledge-based economy."

The failure to regard women's social and communicative skills as knowledge-based and reward them accordingly has strong echoes with the way in which women have been traditionally defined as technically unskilled, thus excluding them from well-paid work. The association between technology, masculinity, and the very notion of what constitutes skilled work was and is still fundamental to the way in which the gender division of labor is reproduced. Further, men's traditional monopoly of technology has been identified as key to maintaining the definition of skilled work as men's work. Machine-related skills and physical strength are basic measures of masculine status and self-esteem, and by implication, the least technical jobs are suitable for women. Technological innovation in the print and newspaper industry provides a clear illustration (Cockburn, 1983). For compositors, the move to computerized typesetting technologies was experienced not only as a threat to their status as skilled workers but also as an affront to their masculinity, and they resisted vigorously. The result is that machinery is literally designed by men with men in mind—the masculinity of the technology becomes embedded in the technology itself (see also Cockburn & Ormrod, 1993; Wajcman, 1991).

A recent report about women in the information technology, electronics, and communications (ITEC) sector confirms that women are making few inroads into technology-related courses and careers (Millar & Jagger, 2001). The report, which covers six countries, including the United States, found that women are generally under-represented among graduates in ITEC-related subjects, despite the fact that they form the majority of university graduates overall. In the United States, for example, women were particularly underrepresented among graduates in computer and information science (27 percent) and engineering (16 percent). At the doctoral level, in computer and information science, women represented but 16 percent; in engineering, only 10 percent were women in 1995 (Fox 1999, 2001). Indeed, the number of women graduating in computer and information science declined from 14,966 in 1985–86 to 6,731 in 1996–97.

This bias in women's and girls' educational choices has major repercussions because ITEC employment is graduate intensive. It is reflected in women's low participation in ITEC occupations across the U.S. economy, which declined from 37 percent in 1993 to 28 percent in 2000. Women are relatively well represented in the lower-status ITEC occupations, such as telephone operators, data processing equipment installers and repairers, and communications equipment operators. By contrast, male graduates are heavily concentrated among computer system analysts and scientists, computer science teachers, computer programmers, operations and systems researchers and analysts, and broadcast equipment operators. In all the countries surveyed, women face considerable barriers when they attempt to pursue a professional or managerial career in ITEC. The result is that "women are chronically underrepresented in ITEC jobs that are key to the creation and design of technical systems" (Millar & Jagger, 2001, p. 16). For all the rhetoric about women prospering in the emerging digital economy, the gender gap is only being slowly eroded.

The multiple causes are all too familiar. Schooling, youth cultures, the family, and the mass media all transmit meanings and values that identify masculinity with machines and technological competence. An extensive literature now exists on sex stereotyping in general in schools, particularly on the processes by which girls and boys are channeled into different subjects in secondary and higher education, and the link between education and gender divisions in the labor market (see Light & Littleton, 1999). Furthermore, many children's computer games and educational software (such as the ubiquitous Game Boy by Nintendo) are clearly coded as "toys for boys." Many of the most popular games are simply programmed versions of traditionally male noncomputer games, involving shooting, blowing up, speeding, or zapping in some way or another. They often have militaristic titles such as "Destroy All Subs" and "Space Wars," highlighting their themes of adventure and violence. No wonder, then, that these games often frustrate or bore the nonmacho players exposed to them. As a result, macho males often have a positive first experience with the computer; other males and most females have a negative initial experience.

The dominant image of the ICT worker is young, white, male "nerds" or "hackers" who work sixteen-hour days and who neither seek nor have access to family-friendly work practices such as part-time and flexible work. Indeed, it is rare to see a female face among the new dot-com millionaires. The "cyber-brat pack" for the new millennium—those wealthy and entrepreneurial young guns of the Internet—consists almost entirely of men. The masculine workplace culture of passionate virtuosity, typified by the hacker-style work, has been well described by Turkle (1984, p. 216) in a chapter entitled "Loving the Machine for Itself." Based on ethnographic research at MIT, Turkle describes the world of computer hackers as epitomizing the male culture of "mastery, individualism, nonsensuality." Being in an intimate relationship with a computer is also a substitute for, and refuge from, the much more uncertain and complex relationships that characterize social life. Turkle's account resonates with Hacker's (1981) studies of engineering where mastery over technology is a source of both pleasure and power for the predominantly male profession.

This is not to imply that there is a single form of masculinity or one form of technology. Rather, it is to note that in contemporary Western society, the culturally dominant form of masculinity is still strongly associated with technical prowess and power. Feminine identity, on the other hand, involves being ill-suited to technological pursuits. Indeed, the construction of women as different from men is a key mechanism whereby male power in the workplace is maintained. A successful career in ICT requires navigation of multiple male cultures associated not only with technological work but also, as we shall see below, with managerial positions.

Feminization of Management?

Compared with their generally unsteady progress in ICT employment, women have made great strides into traditionally male managerial careers, as indicated earlier (Jacobs, 1995). This has given rise to a new orthodoxy that management culture, structures, and practices, which used to be deeply masculine and hostile to women, are being feminized. A growing number of organizational theorists and management consultants assert that, in the new service economy, there is a premium on less hierarchical, more empathetic and cooperative styles of management. According to the new orthodoxy, effective management needs a softer edge, a more qualitative approach (Applebaum & Batt, 1994; Handy, 2001; Kanter, 1989). Successful firms are described as people-oriented and decentralized, uncluttered by bureaucratic layers of management. Leadership is now concerned with fostering shared visions, shared values, shared directions, and shared responsibility. It is suggested that women have a more consensual style of management and thus are ideally suited to postindustrial corporations. The conclusion drawn in this literature is that the norm of effective management will be based on the way women do things.

As in the literature on service work, the management literature focuses on the advantages women have because of communication and social skills, considered to be natural attributes. Reflecting on the copious literature on leadership style, I am struck by the way in which it is permeated by stereotypical dualisms, such as that between hard and soft, reason and intuition (Fagenson, 1993; Helgesen, 1990; Powell, 1993). Instead of challenging the gendered nature of these dichotomies, they are simply inverted. Leadership traits that correspond with male traits, like dominance, aggressiveness, and rationality, are now presented negatively, while formerly devalued feminine qualities, like the soft and emotional, are presented positively.

My own research on senior managers in five high-technology multinational companies sets out to explore these claims; in particular, it explores whether there really is a difference in management style between men and women and, second, how women and men managers are faring in the new management cultures. The study combined a large-scale survey with detailed interviews of male and female managers, and featured a leading American computer corporation. (For a full description of the study, see Wajcman, 1998.)

Initially, I found that a high proportion of both women and men expressed the view that sex differences in management style do exist. On the whole they described women's difference in positive terms. Typical descriptions by both men and women of the male style include: "directive," "self-centred/self-interested," "decisive," "aggressive," and "task oriented." Adjectives used to describe the female style are: "participative," "collaborative," "cooperative," "people-oriented," and "caring." However, when respondents described *their own* management style, either as "participative style, people-handling skills, developing subordinates" or as "leading from the front, drive, decisively directing subordinates," there was no significant difference between the men's and women's responses.

These findings confirm the extent to which people characterize themselves in terms of dominant cultural values. Research shows that men and women tend to stereotype their own behavior according to learned ideas of gender-appropriate behavior, just as they stereotype the behavior of other groups (Epstein, 1988). An integral part of the identity of men and women is the perception that they possess, respectively, masculine and feminine qualities. So it is not surprising that women and men respondents subscribe to gender stereotypes of management styles.

It is only when asked to discuss more specifically their own work practices that the gap between beliefs and behavior emerges. The evidence from the qualitative case-study material reveals a major discrepancy between the current discourse of "soft," people-centered management and the "hard" reality of practice. Many of the interviewees commented that with the almost continuous downsizing of companies, management is returning to a more traditional hierarchical structure. Macho management is again in the ascendancy. Coping with uncertainty within the organization was a constant theme. All respondents found the process of making people redundant difficult, and no obvious gender differences in managerial style emerged in how they accomplished the task. Likewise, when people talked about how they handled other sensitive or conflictual issues, no gender differences emerged. Rather, both men and women told similar stories about how they dealt with such issues as pay, performance, and retrenchment.

The business context of continuous restructuring and job losses has greatly intensified pressures for senior managers. The traditional career-for-life model, based on employment security and promotion prospects, has been replaced by the logic of survival resulting in heightened individualistic competition for a dwindling number of career opportunities. In today's harsh economic climate, *both* men and women feel the need to conform to the "hard" macho stereotype of management because it is still, in practice, the only one regarded as effective.

Both the men and women in my study inhabit a male-dominated, high-tech, corporate world. Managers of both sexes must present themselves so as to project an image of the authoritative manager, adopting or adapting the hegemonic masculine mode. But while they may have the heart for it, women do not have the body for it. The employment contract grants employers command over the bodies of their employees, but these bodies are sexually differentiated. As several organization theorists have noted, male sexuality underpins the patriarchal culture of professional and organizational life (Cockburn, 1991; Gherardi, 1995; Hearn et al., 1989). The sexualization of women's bodies presents a particular problem for women managers.

Whereas writers on the service sector have emphasized the extent to which the sexual skills or services of employees are incorporated into their organizational role, in management, it is men's bodies that are inscribed in the managerial function and women's bodies that are excluded. As Acker (1990, p. 152) has emphasized, women's bodies are often ruled out of order: "women's bodies—female sexuality, their ability to procreate and their pregnancy, breastfeeding, and child care, menstruation, and mythic 'emotionality'—are suspect, stigmatized, and used as grounds for control and exclusion." Without constant vigilance regarding gender self-presentation at work, women run the risk of not being treated seriously as managers. I found that the women in my study had to abandon aspects of their femininity and develop attributes that resemble those of male executives. Far from witnessing a feminization of senior managerial work, my study concludes that women must generally "manage like a man" to succeed in their careers.

Within men's studies, too, there is a growing interest in the masculinity of managers (Cheng, 1996; Collinson & Hearn, 1996; Weiss, 1990) and whether it is being reshaped in the digital economy. Kerfoot and Knights (1996), for example, purport a shift away from the standard form of masculinity associated with modern management practice that is abstract, rational, highly instrumental, future-oriented, strategic, and wholly disembodied. This traditional and paternalistic masculinity is allegedly being displaced by a form of masculinity that displays entrepreneurialism and risk taking. A more egalitarian form of masculinity, it is consistent with the new management approach's emphasis on social relationships. Further feminization of managerial employment is promised, given the association of femininity with interpersonal skills. This claim is contested by Calas and Smircich (1993) who predict that more junior managerial positions, confined to national-level concerns, will continue to be feminized and downgraded while men colonize the more powerful and prestigious globalized functions. Multinational corporations are still run by men and the appointment of women to senior positions is still seen as a risky proposition.

Because of the enormous power held by senior managers and executives, gender inequalities in access to authority constitutes a key mechanism sustaining gender inequalities in the economy at large. For men, the transition from technical roles to managerial jobs is a relatively straightforward step in their career trajectory, whereas women's careers typically hit the proverbial "glass ceiling." An interesting finding of my research was that significantly more women than men reached senior positions via professional credentials. This is consistent with Savage's study, which also found that "women have moved into positions of high *expertise,* but not positions of high *authority*" (Savage, 1992, p. 124; see also Wright & Baxter, 1995). It appears that managerial power and authority are even more intrinsically gendered than technical expertise. While women have benefited enormously from the growth of professional occupations in the information society, men continue to monopolize the elite levels of corporate careers.

ICTs and Changing Location

All visions of the information society place great emphasis on the way ICTs allow for an increasing disassociation between spatial proximity and the performance of paid work (Castells, 1996; Sassen, 1996). The idea is that with the advent of technological innovations, production no longer requires personnel to be concentrated at the place of work. In this scenario, the home as workspace liberates people from the discipline and alienation of industrial production. Computer-based homework or telework offers the freedom of self-regulated work and a reintegration of work and personal life. Moreover, an expansion of teleworking will allegedly lead to much greater sharing of paid work and housework, as men and women spend more time at home together. Mothers are particularly seen as the beneficiaries of this development as working from home allows much greater flexibility to combine employment with childcare.

Futurologists commonly assume a dramatic increase in teleworking, but the term itself suffers from a lack of clarity. If one adopts a narrow definition of teleworkers, as those employed regularly to work online at home, the figures are rather small, perhaps 1–2 percent of the total U.S. labor force (Castells, 1996, p. 395). As many commentators like to joke, more people are researching teleworking than are actually doing teleworking. Nevertheless, teleworking has important implications for the way women's work is understood. We need to distinguish between skilled or professional workers who work from home and the more traditional "homeworkers" who tend to be semi-skilled or unskilled low-paid workers. The former certainly do have more choices about how they schedule their work to fit in with the rest of their everyday lives. However, these teleworkers, who tend to work in occupations like computing and consultancy, are typically men. Women who telework are mainly secretarial and administrative workers. So a rather conventional pattern of occupational sex segregation is being reproduced in this new form of work (Huws et al., 1996).

Indeed, women and men are propelled into teleworking for very different reasons. While women's main motivations are childcare and domestic responsibilities, men express strong preferences for the flexibility, enhanced productivity, convenience, and autonomy of such working patterns. The common media image of a woman working while the baby happily crawls across a computer is misleading. There is an important difference between being at home and being available for childcare. Women continue to carry the bulk of responsibility for domestic work and childcare and, for them, telework does not eliminate their double burden. Even among the minority of professional women who work from home, few are able to separate the demands of motherhood and domesticity from paid work (Adam & Green, 1998). For men, who can more easily set up child-free dedicated "offices" at home, telework often leads to very long and unsocial hours of work. These long hours tend to militate against a more egalitarian and child-centered way of life (Felstead & Jewson, 2000).

More significant than "pure" telework (although it has received much less attention) is the capacity of ICTs to facilitate and encourage people to bring work home from the office. Not only has there been a general increase in the number of hours worked at the workplace for managers and professionals, but also the expectation of availability has been greatly extended with the advent of mobile phones, e-mail, and fax machines. In this sense, the boundaries between the public world of work and the private home have become blurred. However, this is almost always in such a way as to facilitate the transfer of work into the home rather than the transfer of home concerns into the workplace. This has made the balance between work and home at senior levels even more difficult. ICTs may have raised productivity, but they have certainly not reduced working time (Schor, 1991).

Interestingly, here we have echoes of the earlier debate about how domestic technologies would save housewives time. Vacuum cleaners, refrigerators, electric ranges, and washing machines—to most Americans these are mundane features of everyday life. Yet, within the memory of many people living today, they were the leading edge of a revolutionary new laborsaving technology that promised to liberate women from a life of domestic drudgery. Although few would dispute that household technologies have made life easier for millions of women (and men), the average woman who is not employed outside of the home devotes just about the same amount of time to taking care of the household as did her mother and grandmother. Even for women who are employed full-time outside the home, increasing evidence casts doubt on there being a direct relationship between the ownership of household appliances and a reduction in the amount of labor required to accomplish routine housework. Rather than simply saving time, technology changes the nature and meaning of the tasks themselves, which can result in even "more work for mother" (Cowan, 1983; Bittman & Wajcman, 2000).

Any discussion of the geographical relocation of worksites must address the use of female labor in the developing world. Although the international division of labor is not a new phenomenon, innovations in ICTs allow a spatial flexibility for a growing range of tasks. The use of third-world female labor by multinational manufacturing industries offering poorly paid assembly jobs is well documented (Mitter & Rowbotham, 1995; Horton, 1996). Garment assembly and seamstress work is subcontracted to small, offshore companies in the South, while the process of design and cutting is carried out in the North.

However, with increasingly automated means of coordinating marketing, production, and customer demand on a daily basis, garment companies have begun to reverse their reliance on third-world labor. Western and Japanese companies alike are increasingly intent upon "close-to-market" strategies that involve subcontracting work to smaller companies in the West. In this move back to their host countries, companies mainly employ women from immigrant and ethnic minority groups, ensuring a captive, regional labor force, compelled to accept low wages and exploitative working conditions. Despite the high levels of capital investment and advanced ICTs used by these firms, there is little transfer of technical skills and expertise to the women who work in these manufacturing jobs.

White-collar, professional, and clerical jobs are also moving to developing countries, underpinned by the diffusion of ICTs. The emergence of the Indian information technology industry is a notable case. Skilled software design and development projects are sited in the West, whereas the programmers employed are located in offshore companies, maintaining consistently high production rates for very low wages. Countries like India have a ready pool of female labor available for software work—women who are well educated, English speaking, and technically proficient. It is estimated that, in India, women constitute over 20 percent of the total IT workforce, which is higher than women's participation in the Indian economy as a whole (Kelkar & Nathan, 2002). Despite the repetitive nature of the work and the lack of job security, women's income, authority in household matters, and social mobility have improved as a result.

Alongside the export of software jobs to countries such as India, Mexico, and China, a proliferation of cybercafes and computers can provide the means for women's groups to organize networks and campaigns to improve their conditions. Indeed, the combination of information technology with telecommunications, particularly satellite communications, provides new opportunities and outlets for women. In country after country, although women still account for a lower proportion of Internet users, and may use the Internet for different purposes than men do, their share is rapidly rising (Sassen, 2002). The fear that the globalization of communications would lead to homogenization and reduce sociability and engagement with one's community was ill conceived. On the contrary, new electronic media have helped build local communities and project them globally. Cyberspace makes it possible for even small and poorly resourced nongovernmental organizations to connect with each other and engage in global social efforts. These political activities are an enormous advance for women who were formerly isolated from larger public spheres and cross-national social initiatives. "We see here the potential transformation of women, 'confined' to domestic roles, who can emerge as key actors in global networks without having to leave their work and roles in their communities" (Sassen, 2002, p. 380). This is not to endorse utopian ideas of cyberspace being gender-free and the key to women's liberation. Rather, it is to stress that the Internet, like other technologies, contains contradictory possibilities and can be a powerful tool for feminist politics.

Conclusion

We need to keep a skeptical eye on purely technological interpretations of the effects of digital technology. Moreover, we need to recognize the embeddedness of social relations and the variable, sometimes contradictory, outcomes of these technologies for different groups of women. ICTs can be constitutive of new gender power dynamics, but they can also be derivative of or reproduce preexisting conditions of gender inequality at work. While the rise of the service economy has lead to a feminization of the labor force, these new forms of work to some extent replicate old patterns of sex segregation. The skills that are exercised in predominantly female jobs are still undervalued, and women are slowly making inroads into the upper echelons of ICT occupations. This is directly related to the enduring problem of women's exclusion from senior levels of management. While the flexibility and spatial mobility afforded by ICTs have expanded opportunities for women, the male cultures associated with technical and managerial expertise still serve as a brake on progress toward gender equality at work.

References

Acker, J. (1990). Hierarchies, jobs, bodies: A theory of gendered organizations. *Gender and Society, 4*, 139–158.

Adam, A., & Green, E. (1998). Gender, agency and location and the new information society. In B. Loader (Ed.), *Cyberspace Divide* (pp. 83–97). London: Routledge.

Adkins, L. (1995). *Gendered work: Sexuality, family and the labor market.* Buckingham, Eng.: Open University Press.

Applebaum, E., & Batt, R. (1994). *The new American workplace.* New York: ILR Press.

Bell, D. (1973). *The coming of post-industrial society.* New York: Basic Books.

Bittman, M., & Wajcman, J. (2000). The rush hour: The character of leisure time and gender equity. *Social Forces, 79*(1), 165–189.

Bradley, H., et al. (2000). *Myths at work.* Cambridge: Polity Press.

Burris, B. H. (1998). Computerisation of the workplace. *Annual Review of Sociology, 24*, 141–157.

Calas, M., & Smircich, L. (1993). Dangerous liaisons: The "feminine-in-management" meets "globalization." *Business Horizons,* March/April, 71–81.

Castells, M. (1996). *The rise of the network society.* Oxford: Blackwell.

Cheng, C. (Ed.). (1996). *Masculinities in organizations.* London: Sage Publications.

Cockburn, C. (1983). *Brothers: Male dominance and technological change.* London: Pluto Press.

———. (1991). *In the way of women: Men's resistance to sex equality in organizations.* London: Macmillan.

Cockburn, C., & Ormrod, S. (1993). *Gender and technology in the making.* London: Sage Publications.

Collinson, D., & Hearn, J. (Eds.) (1996). *Men as managers, managers as men.* London: Sage Publications.

Cowan, R. S. (1983). *More work for mother.* New York: Basic Books.

Du Gay, P. (1996). *Consumption and identity at work.* London: Sage Publications.

Epstein, C. Fuchs (1988). *Deceptive distinctions: Sex, gender, and the social order.* New Haven, Conn.: Yale University Press.

Fagenson, E. (Ed.) (1993). *Women in management: Trends, issues, and challenges in managerial diversity.* Newbury Park: Sage Publications.

Felstead, A., & Jewson, N. (2000). *In work, at home: Towards an understanding of homeworking.* London: Routledge.

Forrest, A. (1993). Women and industrial relations theory. *Relations Industrielles, 48,* 409–440.

Fox, Mary Frank. (1999). Gender, hierarchy, and science. In J. S. Chafetz (Ed.), *Handbook of the sociology of gender* (pp. 441–457). New York: Kluwer Academic.

———. (2001). Women, men, and engineering. In D. Vannoy (Ed.), *Gender mosaics* (pp. 249–257). California: Roxbury.

Gherardi, S. (1995). *Gender, symbolism and organizational cultures.* London: Sage Publications.

Hacker, S. (1981). The culture of engineering. *Women's Studies International Quarterly, 4,* 341–53.

Handy, C. (2001). *The elephant and the flea: Looking backwards to the future of work.* London: Hutchinson.

Hearn, J., et al. (Eds.). (1989). *The sexuality of organization.* London: Sage Publications.

Held, D., McGrew, A., Goldblatt, D., & Perraton, J. (1999). *Global transformations: Politics, economics and culture.* Cambridge: Polity Press.

Helgesen, S. (1990). *The female advantage: Women's ways of leadership.* New York: Doubleday.

Hochschild, A. (1983). *The managed heart: Commercialization of human feeling.* Berkeley: University of California Press.

Horton, S. (Ed.). (1996). *Women and industrialization in Asia.* London: Routledge.

Huws, U., et al. (1996). *Teleworking and gender* (Report 317). Brighton, U.K.: Institute for Employment Studies.

Jacobs, J. (Ed.). (1995). *Gender inequality at work.* Thousand Oaks, Calif.: Sage Publications.

Jasanoff, S., et al. (Eds.). (1995). *Handbook of science and technology studies.* Thousand Oaks, Calif.: Sage Publications.

Kanter, R.M. (1989). *When giants learn to dance: Mastering the challenge of strategy, management and careers in the 1990s.* New York: Simon & Schuster.

Kelkar, G., & Nathan, D. (2002). Gender relations and technological change in Asia. *Current Sociology, 50*(3), 427–441.

Kerfoot, D., & Knights, D. (1996). "The best is yet to come?" The quest for embodiment in managerial work. In D. Collinson & J. Hearn (Eds.), *Men as managers, managers as men* (pp. 78–98). London: Sage Publications.

Kumar, K. (1995). *From post-industrial to post-modern society: New theories of the contemporary world.* Oxford: Blackwell.

Leidner, R. L. (1993). *Fast food, fast talk: Service work and the routinization of everyday life.* Berkeley: University of California Press.

Light, P., & Littleton, K. (1999). *Social processes in children's learning.* Cambridge, U.K.: CUP.

Loader, B. (Ed). (1998). *Cyberspace divide.* London: Routledge.

Mackenzie, D., & Wajcman, J. (1999). *The social shaping of technology* (2nd ed.). Milton Keynes, U. K.: Open University Press.

Milkman, R. (1995). Economic inequality among women. *British Journal of Industrial Relations 33*(4), 679–683.

Millar, J., & Jagger, N. (2001). *Women in ITEC courses and careers.* London: Women and Equality Unit, U.K. Department of Trade and Industry.

Mitter, S., & Rowbotham, S. (Eds.). (1995). *Women encounter technology.* London: Routledge.

Powell, G. (1993). *Women and men in management.* Newbury Park, Calif.: Sage Publications.

Rubery, J., et al. (1998). *Women and European employment.* London: Routledge.

Sassen, S. (1996). *Losing control? Sovereignty in an age of globalization.* New York: Columbia University Press.

———. (2002). Towards a sociology of information technology. *Current Sociology, 50*(3), 365–388.

Savage, M. (1992). Women's expertise, men's authority: Gendered organisations and the contemporary middle classes. In M. Savage & A. Witz (Eds.), *Gender and bureaucracy.* Oxford, U.K.: Blackwells.

Schor, J. (1991). *The overworked American: The unexpected decline of leisure.* New York: Basic Books.

Turkle, S. (1984). *The second self: Computers and the human spirit.* London: Granada.

Tyler, M., & Abbott, P. (1998). Chocs away: Weight watching in the contemporary airline industry. *Sociology, 32*, 433–450.

Wajcman, J. (1991). *Feminism confronts technology.* University Park: Pennsylvania State University Press.

———. (1998). *Managing like a man: Women and men in corporate management.* University Park: Pennsylvania State University Press.

———. (2004). *TechnoFeminism.* Cambridge, U.K.: Polity Press.

Webster, F. (1995). *Theories of the information society.* London: Routledge.

Webster, J. (1999). Technological work and women's prospects in the knowledge economy. *Information, Communication & Society, 2*(2), 201–221.

———. (2000) Today's second sex and tomorrow's first? Women and work in the European information society. In K. Ducatel, J. Webster, & W. Herrmann (Eds.), *The information society in Europe* (pp. 119–140). Lanham, Md.: Rowman & Littlefield.

Weiss, R. S. (1990). *Staying the course: The emotional and social lives of men who do well at work.* New York: Free Press.

Wright, E. O., & Baxter, J. (1995). The gender gap in workplace authority: A cross-national study. *American Sociological Review, 60*, 407–435.

Yeuk-Mui Tam, M. (2001). Information technology in frontline service work organization. *Journal of Sociology, 37*(2), 177–206.

Zuboff, S. (1988). *In the age of the smart machine.* New York: Basic Books.

5

Gender, Race/Ethnicity, and the Digital Divide

CHERYL B. LEGGON

One major issue on the United States' science and technology policy agenda since the late 1980s has been the digital divide—the growing gap between those with access to telephones, modems, computers, and the Internet, and those without such access: the information-rich versus the information-poor.

This issue provides an excellent opportunity to highlight the crucial importance of examining the effects of race/ethnicity and gender in the use of technology. This chapter seeks to do so by focusing on the divide by race/ethnicity and gender with regard to access to and use of information technology.

There is consensus about the proliferation of scientific and technical information: this information almost doubles every five years; however, there is dissensus about the potential impact of information technology. Some contend that information technology can close (or significantly assist in closing) gaps between groups insofar as it facilitates all groups' having voice in the public arena (President's Information Technology Advisory Committee [PITAC], 1999). Others argue that information technology further widens the gap (Tapscott, 1998). With information, as with other resources, both advantage and disadvantage are cumulative.

The divide is more than an issue of access to technology; it is also an issue of use and empowerment. Some researchers contend that use of the Internet negatively affects the formation and maintenance of social networks. Others argue that this technology facilitates the development of support networks (Educom Review, 1999). Discussions of this divide often focus on differences in terms of access by race/ethnicity *or* gender.[1] Although this dichotomous categorization (either race/ethnicity or gender) reflects the categories by which data on the U.S. science and technical (S/T) workforce have been traditionally collected and presented, it is still problematic for at least two reasons. First, the dichotomy implies that the term "minorities" refers to minority males, and the terms "gender" and "women" refer to white women. This relegates minority women and/or women of color to footnotes or endnotes in discussions of both minorities and women—if they are discussed at all. Second, it fails to examine the intersection of gender with race/ethnicity.

This chapter seeks to enhance understanding of the intersection of race/ethnicity *and* gender in the production and use of information technology. The focus is on African American and Hispanic women because these women are part of the two largest "minority" groups in the United States. According to the 2000 census, Hispanics represent 12.5 percent of the U.S. population, while African Americans represent 12.3 percent (U.S. Census Bureau, 2000). This chapter defines African American women as being born and raised in the United States and having African ancestry. Hispanic women are defined as belonging to the following ethnic groups: Mexican American, Puerto Rican, Cuban, Central and South American, and other Hispanic.[2] "Hispanic" is an umbrella term that, unfortunately, obscures significant differences among women who are Mexican American, Puerto Rican, and Cuban. There is dissensus among members of these groups about whether to use the term "Hispanic" or "Latina." Some contend that "Latina" emphasizes the central and South American origins—including Indian heritage—while the term "Hispanic" obscures those origins (Darlington & Mulvaney, 2003). This chapter uses the term "Hispanic" only because it is used by the U.S. Census Bureau.

Some data on Hispanics are collected in such a way that they fail to make at least two important distinctions: among the subgroups of Hispanics—Mexican Americans, Puerto Ricans, Cubans, Central and South Americans, and other Hispanics;[3] and between Puerto Ricans who grew up and were educated on the island versus those who grew up and were educated on the mainland.

This chapter examines Hispanic and African American women not only as users but also as producers of information technology. This examination includes the following: access and orientation to technology; trends in degrees earned in computer science and technology; and trends in labor force participation rates in information technology fields. The approach identifies significant differences in relationships to information technology between black and Hispanic women on the one hand, and their male race/ethnic counterparts and white gender counterparts on the other. The chapter concludes with a discussion of policy implications.

"Information technology" is a term for which there is no clear and consistent definition. For the purposes of this discussion, however, I will use Freeman and Aspray's (1999) definition: "information technology (IT) refers only to computer-based systems. It includes computer hardware and software, as well as the peripheral devices most closely associated with computer-based systems" (p. 25). Computer-based systems are defined to include "the full gamut of technological considerations, ranging from the design and production of chips, through the design and creation of complex, computer-based systems for a particular application, to the end-use of such systems" (p. 35).

African American and Hispanic Women as Consumers

COMPUTER OWNERSHIP

Between 1994 and 1998, there were increases in the percentages of households with a computer among both African American and non-Hispanic white households, although the percentage increase was greater for non-Hispanic whites than for African Americans (National Telecommunications and Information Administration [NTIA], 1999). Both African American and Hispanic households were twice as likely to own computers in 1998 as they were in 1994. Nevertheless, Hispanic households are still only half as likely as non-Hispanic white households to own a computer. At the highest income levels ($75,000 and above), the non-Hispanic white/African American divide for computer ownership decreased by 76.2 percent between 1994 and 1998. In fact, the higher incomes are not only helping to narrow the gap, but they also could even be closing it.

INTERNET ACCESS

"Internet access is no longer a luxury item, but a resource" (NTIA, 2000, p. 3). In the United States, Internet access is affected by race/ethnicity, geography, income level, and education level, and it tends to be greater in urban than in rural areas, and highest in the central cities. This pattern applies across all race and ethnic groups. Between 1998 and 2000, Internet access increased 110 percent among black households, 87 percent among Hispanic households (NTIA, 2000). However, it is important to note that groups with lower rates of access exhibit higher expansion rates because they are starting from a much lower base and have greater opportunity for rapid and broad expansion (NTIA, 2000, p. 19). Similarly, between 1998 and 2000, expansion rates were highest for those at lower levels of education.

The digital divide continues to increase between non-Hispanic whites, on the one hand, and Hispanics and African Americans on the other hand in terms of comparing African Americans and Hispanics both to the national average and to non-Hispanic whites. This divide increases—except at the highest income levels ($75,000 and above). The Internet divide between access rates for African American households and the national average in 2000 was 18 points; this is a 3-point increase since 1998. Similarly, in 2000, the Internet divide between Hispanic households and the national average was 18 points; this was 4 points wider than in 1998. When holding income constant, African American and Hispanic households are still much less likely to have Internet access than are non-Hispanic white households. At the highest income levels ($75,000 and above), the non-Hispanic white/African American divide for computer ownership decreased by 76.2 percent between 1994 and 1998. The high rates of expansion for African Americans and Hispanics suggest that this widening will subside (NTIA, 2000). Nevertheless, it is important to note that 70 percent of African Americans and Hispanics are not connected to the Internet. Moreover, "blacks and Hispanics, particularly poor blacks and Hispanics, are still far behind non-Hispanic white and Asian Americans in access to computers, particularly at home" (Digital Divide Network, 2002). While the digital divide in terms of computer access is closing, some observers claim that broadband connectivity "threatens to become the next great digital divide" (Digital Divide Network, 2002).

Both computer ownership and Internet access and use are strongly influenced by income and education. Some evidence suggests that both education and income are independently associated with Internet access. However, controlling for education and income accounts for only half of the gap in Internet usage between African Americans and Hispanics and the national average, as shown in Figure 5.1.

Note that Figure 5.1 shows two levels of racial/ethnic gap in Internet access gap: the actual levels of gap, and levels of gap adjusted for income and education.

INTERNET USAGE

Focusing on individuals reveals important differences in Internet use based on age, gender, and labor force status (NTIA, 2000, p. 12). In December 1998, 32.7 percent of the U.S. population used the Internet; that number increased to 44.4 percent in August 2000 (p. 51). In 1998, 34.2 percent of men and 31.4

Figure 5.1. Gaps in U.S.* Internet Access by Race/Ethnicity for 1998 and 2000: Actual and Adjusted**

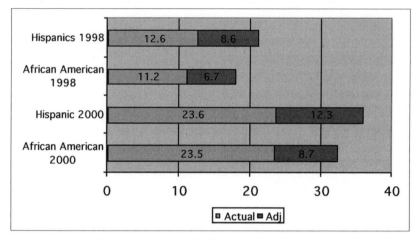

Source: National Telecommunications and Information Administration (NTIA), 2000, Figures 1–12. Data from NTIA, ESA, U.S. Department of Commerce using U.S. Bureau of the Census Current Population Supplements.

*Figures for the nation represent the percentage with Internet access; this is the group upon which both the actual and adjusted gaps are calculated.

**Adjusted for education and income.

percent of women used the Internet; by 2000, these percentages increased to 44.6 percent and 44.2 percent respectively—not statistically different.[4]

Overall, gender differences in Internet usage have almost disappeared. As shown in Figure 5.2, very small gender differences remain within some race and ethnic groups. The usage for non-Hispanic white women and men appears to be identical. For both African Americans and Hispanics, only 2–3 percent more women than men use the Internet; however, among Asian Americans, about 5 percent more men than women use the Internet.

As shown in Table 5.1, both Hispanic women and men most frequently use the Internet to send and read e-mail and to research products. Hispanic women use the Internet more frequently than Hispanic men for getting health information and instant messaging. Hispanic men use the Internet more frequently than Hispanic women for getting news and researching products.

Like their Hispanic counterparts, African American women and African American men use the Internet most frequently to send and read e-mail. African American women (like Hispanic women) are more likely to use the Internet more frequently than African American men to find health information, as shown in Table 5.2.

Figure 5.2. Internet Use by Gender and Race/Ethnicity, 2000

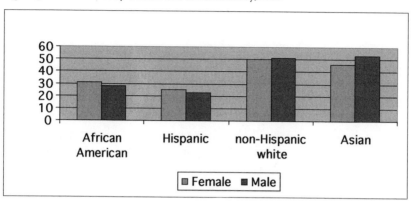

Source: National Telecommunications and Information Administration (NTIA), 2000

Table 5.1: Gender Differences in Types of Internet Usage among Hispanics, 2000

	Females	Males
Send and read e-mail	88%	83%
Get news	56%	65%
Research products	63%	81%
Find health info.	59%	42%
Use instant messaging	55%	46%

Margin of error is +−1%.
Source: PEW Internet and American Life Project, 2000 Tracking Survey.
N = 26,094.

Table 5.2: Gender Differences in Types of Internet Usage among African Americans, 2000

	Females	Males
Send and read e-mail	90%	86%
Get news	60%	66%
Research products	75%	73%
Find health info.	62%	49%
Use instant messaging	46%	55%
Margin of error is +−1%.		

Source: PEW Internet and American Life Project, 2000 Tracking Survey.
N = 26,094.

Hispanic and African American Women as Producers of Information Technologies

Neither the concept "information technology" nor the concept "information technology worker" has a precise definition. Depending on how information technology is defined, there are as many as twenty academic specialties that study "various aspects of its use and applications" (Freeman & Aspray, 1999, p. 28). For the purposes of this chapter, numbers of academic degrees in three of these disciplines will be used as a proxy for information technology workers. The three disciplines are computer science, computer engineering, and information systems. The numbers of degrees at various levels will be used as proxies for estimating the composition of the IT workforce by race/ethnicity and gender.[5]

Research indicates low numbers of both women and minorities in IT careers (Virginia Tech Forum, 1999). In 1999 in the United States, women of color represented 3 percent of all computer/mathematical scientists employed in the workforce, while men of color represented 4.9 percent of such workers. Also in that same year, women of color represented 2.8 percent of all computer/information scientists in the United States workforce, while men of color comprised 4.9 percent (National Science Foundation [NSF], 1999). There is an inverse correlation between degree level and numbers of degrees awarded to African American and Hispanic women.

BACHELOR'S DEGREES

Figure 5.3 shows the number of bachelor's degrees in computer science awarded to women by race and ethnicity for selected years from 1990 to 2001. For both African American women and Hispanic women, the overall trend has been a gradual increase in the numbers of bachelor's degrees awarded during this period. Both African American and Hispanic females experienced larger percentage increases in the numbers of computer science bachelor's degrees earned than did non-Hispanic white women (44 percent). However, it is important to note that the 1990 baseline numbers for both African American females (1,173) and Hispanic females (416) were much smaller than those for non-Hispanic white women (4,944) (NSF, 2003, Table 3–15). For Hispanic women, the number of degrees in computer science fluctuates. For African American women, numbers of bachelor's degrees awarded in computer science increased each year from 1992 to 1995 and from 1997 to 2001.

Among non-Hispanic white women, bachelor's degrees awarded in computer science between 1990 and 2001 decrease each year from 1990 to 1997 and

Figure 5.3. Computer Science Bachelor's Degrees Awarded to Women by Race/Ethnicity for Selected Years, 1990–2001

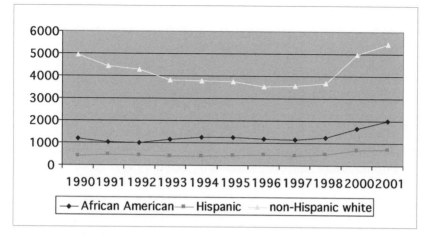

Source: National Science Foundation (2003). *Women, Minorities, and Persons with Disabilities in Science and Engineering* (NSF 03-312). Arlington, Va: Appendix table 3-15; National Science Foundation (2004). *Women, Minoritie, and Persons with Disabilities in Science and Engineering* (NSF 04-317). Arlington, Va: Table C-13.

then increase each year from 1998 to 2001 (NSF, 2003, Table C-13; NSF 2003, Table 3–15). Between 1990 and 2001, the percentages of bachelor's degrees in computer science awarded to both African American women and Hispanic women increased slightly as a percentage of the total computer science degrees awarded, while the percentage awarded to non-Hispanic white women decreased (NSF 2004, Table C-13; NSF 2003, Table 3-15).

It is interesting to note that the top academic institutions awarding bachelor's degrees in computer science to both African American and Hispanic women are minority-serving institutions (MSIs), which consist of historically black colleges and universities (HBCUs), Hispanic-serving institutions (HSIs), and tribal colleges and universities (TCUs) (NSF Web-CASPAR Database System).

Similarly, HSIs awarded the largest number to Hispanic women. This has important implications for policy in terms of increasing the numbers of computer science degrees earned by African American women and Hispanic women.

One implication is that since these institutions that successfully develop computer science talent have been identified, funds can be targeted to sustain and enhance to an even greater extent their computer science degree training for African American women and Hispanic women. Another implication is that these institutions could develop ongoing "feeder" relationships with doctorate-granting institutions in order to help African American female and Hispanic female students earn master's degrees and doctorates in computer science.

MASTER'S DEGREES

Despite periods of decline—1992–93, 1994–95, and 1997–98—there is an overall upward trend in the number of computer science master's degrees earned by African American women between 1990 and 1998. During the same period, Hispanic women experienced only one period of decline (1993–94), a steady state from 1994–96, and then an upward trend from 1997–98.

For 1997–98, both African American and Hispanic women earned a small proportion of all master's degrees awarded in computer and information science. African American women were awarded 169 such degrees—45 percent of all master's degrees in computer science earned by African Americans, and 10 percent earned by women (excluding those awarded to nonresident aliens). Hispanic women were awarded 51 master's degrees in computer and information science. This represents 27 percent of such degrees awarded to Hispanics, 3 percent of those awarded to all women, and 0.8 percent of the total number of master's degrees awarded. For each year from 1990 to 2000, among African Americans, the percentage of women earning master's degrees in computer science was larger than the percentages for African American men and all women (CPST, 2002, Tables 2-25, 2-26, and 2-27).

DOCTORAL DEGREES

There are no discernible trends in terms of the numbers of doctorates in computer science earned by African American and Hispanic women. The numbers fluctuate from one year to the next. From 1987 to 2000, the *total* number of computer science doctorates earned by African American women was 43; the *total* number for Hispanic women was 27 (Commission on Professionals in Science and Technology [CPST], 2002, Table 6-39). These numbers certainly do not bode well for increasing the numbers of African American and Hispanic females on computer faculties in colleges and universities.

IT Employment in the United States

In the United States, the size of the IT workforce peaked in 2000 and lost 500,000 jobs in 2001. However, by the beginning of 2003, there were 4.2

percent more IT workers than there were at the start of 2002. According to a survey conducted by the Information Technology Association of America (ITAA), the demand for IT workers was at a historic low in 2003 (CPST, 2003). Survey data indicated that the demand for IT workers in 2003 was 493,000 positions over the next year—down from 1.6 million at the start of 2000 and less than half of the predicted 1.1 million positions needed in 2002. The demand continued to drop between 2003 and 2004 (ITAA, 2004).

Figure 5.4 shows the job categories of IT workers. As this figure shows, programmers/software engineers comprise the largest job category; the second largest category is technical support personnel, followed by enterprise systems specialists and database developers/administrators. In 2000, African Americans represented 5.4 percent of all computer programmers and 7.1 percent of computer systems analysts—two of the core jobs in the industry (Hicks, 2002).

Policy Implications

Within the past fifteen to twenty years in the United States, one policy issue cuts across these fields—the digital divide. Although literature on race, gender, and economic equality exist within the interdisciplinary fields of women's studies, African American studies, and Latino/Chicano studies, they have not systematically been part of policy debates.

The data indicate that the digital divide—i.e., the technological gap that exists regarding access to and use of information technology—between Afri-

Figure 5.4. Job Categories of IT Workers as Percent of IT Workforce, 2003

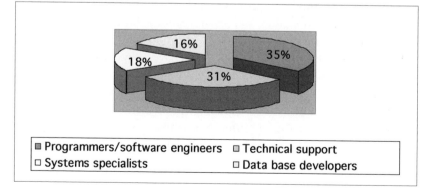

Source: CPST, *Comments,* 40(4) (June 2003).

can American and Hispanic women on the one hand, and non-Hispanic white women on the other, is closing. This narrowing of the gap is due as much to a decrease in the number of computer science bachelor's degrees earned by non-Hispanic white women as to the increase of such degrees earned by African American and Hispanic women.

The narrowing of the digital divide can be partially attributed to concerted efforts on the part of the government at all levels (federal, state, regional, and local) and the private sector. There are programs and activities to increase the recruitment and retention of women in general and women of color into computer science. For example, a program for ninth- and tenth-grade minority girls, funded by the National Science Foundation from 1989 through 1994 at George Washington University, taught computer skills as a means to developing an interest in science and technology. Through this program, professional associations, such as the ITAA, are also working on this issue.

Another factor contributing to the narrowing of the digital divide has to do with how it is presented as a public policy issue. Some support for closing the divide came from arguments that doing so would benefit society as a whole (not only in terms of providing gainful employment for many, but also for enabling more citizens to become technologically literate—and presumably better informed), and to increase productivity and international competitiveness.

Since the 1990s, the digital divide in the United States has been systematically tracked by organizations such as the National Technological Agency and the PEW Charitable Trust in the public and private sectors, respectively. These data have been used to design and implement programs and activities aimed at closing the digital divide.

Initiatives to address this issue must be research driven. This research can come from the interdisciplinary fields of Women's Studies, African American Studies, Chicano Studies, and Science Studies. Ideally, these initiatives should not be "add-ons" but should be institutionalized so that they persist over time. Moreover, these initiatives should have as an integral component systematic evaluation—formative and summative.[6] One critically important aspect of the digital divide concerns the intersection of race/ethnicity and gender. Failure to address this intersection will result in a policy that is flawed at best, and ineffective—even counterproductive—at worst.

Notes

1. For an exception, see Richard J. Coley, *Differences in the gender gap: Comparisons across racial/ethnic groups in education and work*, Princeton, N.J.: Educational Testing Service, 2001.

2. This is consistent with racial and ethnic categories currently used by the National Science Foundation.

3. These are the groups that the U.S. Census Bureau includes in the category "Hispanic."

4. Disaggregating gender data by age reveals that the gender gap in Internet usage widened—in favor of females.

5. This is similar to the approach articulated by Jane Siegel from the Software Engineering Institute in the National Research Council's report *Computing professionals—Changing needs for the 1990s*, Washington, D.C.: National Academy Press.

6. Formative evaluations are conducted throughout the duration of a program or project to determine what is working well and what is not. The data provide real-time feedback to make whatever changes are warranted. Summative evaluations are conducted at the end of a program or project to assess overall efficiency and effectiveness.

References

Browne, Irene. (2000). *Latinas and African American women at work: Race, gender, and economic inequality.* New York: Russell Sage Foundation.

Commission on Professionals in Science and Technology (CPST). (2002). *Professional women and minorities: A total human resources data compendium.* Washington, D.C.: CPST.

CPST. (2003, June). *Comments, 40*(4).

Darlington, Patricia S. E., & Becky Michele Mulvaney. (2003). *Women, power, and ethnicity.* New York: Haworth Press.

Digital Divide Network (2002). *Falling through the net report: A reason for concern and optimism.* Retrieved January 11, 2002, from http://www.digitaldividenetwork.org

Educom Review. (1999). Gary Chapman: Creating E-Community; Important lessons learned from the information technology have-nots. *EDUCAUSE,* January/February. Retrieved August 8, 2001, from http://www.educause.edu/ir/library/html/erm9919.html.

Freeman, Peter, & William Aspray. (1999). *The supply of information technology workers in the United States.* Washington, D.C.: Computing Research Association.

Hicks, Jennifer G. (2002). Unemployment rate remains steady for women. Retrieved December 7, 2002, from http://www.imdiversity.com/villages/woman

Information Technology Association of America (ITAA). (2001, April). When can you start? Rebuilding better information technology skills and careers (an ITAA study).

ITAA. (2004, September). Adding value, growing careers: The employment outlook in today's increasingly competitive IT job market (an ITAA Workforce Development Survey).

National Telecommunications and Information Administration (NTIA). (1999, July). *Fact Sheet: Racial divide continues to grow.* Washington, D.C.: U.S. Department of Commerce.

National Telecommunications and Information Administration (NTIA). (2000). Falling through the net: Toward digital inclusion. Washington, D.C.: U.S. Department of Commerce.

National Science Foundation (NSF). (1999). [Data abstracted from] *Scientists & engineers statistical data system.* Arlington, Va.: Division of the Department of Science Resources Statistics, NSF.

NSF. (2000). *Women, minorities, and persons with disabilities in science and engineering* (NSF00-327). Arlington, Va.: NSF.

NSF. (2001). *Science and engineering degrees, by race/ethnicity of recipients: 1990–98* (NSF01-327). Arlington, Va.: NSF.

NSF. (2003). *Women, minorities, and persons with disabilities in science and engineering* (NSF03-312). Arlington, Va.: NSF.

NSF. (2004). *Women, minorities, and persons with disabilities in science and engineering* (NSF04-317). Arlington, Va.: NSF.

President's Information Technology Advisory Committee (PITAC). (1999, February). *Information technology research: Investing in our future; A report from the President's Information Technology Advisory Committee.* Washington, D.C.: U.S. Department of Commerce.

Tapscott, D. (1998). *Growing up digital: The rise of the net generation.* New York: McGraw Hill.

United States Census Bureau. (2000).

Virginia Tech Forum. (1999). *Women and minorities in information technology forum: Causes and solutions for increasing the numbers in the workforce pipeline; A white paper.* Hampton, Va.: Old Dominion University.

6

Genetic Technology and Women

BARBARA KATZ ROTHMAN

Introduction

My work in the sociology of genetics, which I have been doing for over twenty years now, has led me to a rather depressing conclusion: genetic science, and the technologies that have developed from it, grow out of and maintain traditional gendered divisions. While clinical and reproductive genetics, like much of modern medical management, is used to give an illusion of power and control, it rarely actually empowers women, either as individuals or as a class.

This is not intended as a statement to apply to every use of clinical genetics in every situation. Certainly, some women feel that genetic counseling helped them make better and more informed reproductive decisions, and some feel that genetic counseling helped them with preventative treatment for cancers and other diseases. Looking back on twenty years of research and thought, however, makes me believe strongly that this is neither the genuine reason for our collective investment in genetic science, nor the dominant outcome.

This chapter first lays out the gendered ideology that I believe provides the basis for genetics, both as an ideology and as a science. It then considers the two contexts of procreation and breast cancer—in which new genetics is presented as "empowering" women, or offering women choice—arguing that in neither area has the reason for, or the result of, the work been to empower

Passages of this chapter were drawn from Barbara Katz Rothman, *The book of life: A personal and ethical guide to race, normality, and the implications of the human genome project,* Boston: Beacon Press, 2001.

or better the lives of women. At best, a narrowed understanding of "choice" has offered a few women a few choices in a few situations. "Choice" without power is no real choice at all.

The Underlying Logic of Patriarchy

The deep logic of genetics, the frame or prism of understanding that genetics give us, is that genes are *causes*. Evelyn Fox Keller summarizes the logic of genetic thinking: "genes are the primary agents of life; they are the fundamental units of biological analysis; they cause the development of biological traits; and the ultimate goal of biological science is the understanding of how they act" (Keller, 1995, p. 4). If genes are the cause, the active force, the predictor of traits, then to read genes is to predict traits.

But geneticists often cannot predict traits. There is a murkiness in the translation, rather than the direct cause and effect the logic suggests. Therefore, geneticists have made a useful distinction, noting the difference between genotype and phenotype. The genotype is the genetic reading. The phenotype is how the genotype actually plays out in the being before you. "Environment" becomes that which muddies the waters, that which interferes between the genotype and the phenotype. In a particular pregnancy, for example, the poorly placed identical twin will be smaller than the more felicitously implanted twin; their different environments within the same womb account for their phenotypical difference in spite of their genotypical identity.

This model causes something funny to happen to our thinking. The genotype comes to seem more real somehow, more authentic. The smaller twin was stunted, we tend to think, rather than that the larger being very well nurtured. Because of genetic determinist thinking, the environment—at the level of the cytoplasm of the rest of the cell, outside the DNA-laden nucleus; of the body that carries that embryo; of the community that shelters that maternal body; of the society that contains that community; of the planet on which we all live—all of the environment comes to be seen as that which might interfere with the expression of the authentic genotype.

Genetics—a way of thinking, an ideology as much as a science—places all of the essence of life, all of its energy, majesty, and power, into the nucleus of the cell. The old-fashioned word for that essential bit, that source of life, was the "seed." In the history of Western society, that seed came from men. Our standard answer to "Where babies come from?" is a variation on Daddy plants a seed in Mommy.

Modern thinking recognizes women's seeds and has extended some of the privileges of fatherhood to women. We no longer talk of women as just vessels for the children of men, but recognize women as being connected to their children, as men are, through their seeds. Children, we say, are "half his, half hers"—as if they might as well have grown in the backyard rather than in the bodies of women.

In this seed-based way of thinking, in both its most primitive forms and in the most sophisticated genetics, people talk about a genetic tie, a connection by seed to indicate biological connectedness. We use the word "blood," in older language of bodily connectedness, and talk about "blood ties." But we mean "genetic ties," the seed connection: the one absolutely bloodless part of making babies.

Now that the seeds of women can be separated from our bodies, as the seeds of men can be, we can move them from woman to woman. In the kind of intellectual play that seems fascinating to the bioethicists of reproductive technology, we ask, "If Mary's egg grows to a baby in Susan's body, who is the mother?" We offer two answers to this. One answer is based on who the father is. If Mary's husband has fertilized this egg, then he and Mary are the parents, and Susan merely the "surrogate gestator," the place their baby grows. If Susan's husband has fertilized the egg, then the baby is his and Susan's, and Mary is just the "egg donor." In practical terms, patriarchy and class rule: the man with the checkbook is the father and the woman married to him is the mother.

The other answer we give, though, is a bit more complicated. We—we Americans, we as a society, more and more of "us" (if not me, and maybe not you)—distinguish between the "real" and the apparently not-so-real mother. The "real" mother is most often held to be the woman who supplied the egg. The child is "really" hers. The other woman, surrogate gestator or purchaser of "donated" eggs, is not "really" the mother, but more of an "adoptive" mother, even if the adoption takes place when the "child" does not yet exist as more than a (fertilized) egg in a pipette.

It's a funny reality, that. A woman—great with child, a woman whose baby is drawing its nourishment from her very blood, a woman who is grunting and pushing that baby out of her body, a woman who is attached by the umbilical cord to the baby newly emerged from her vagina—that woman is not "really" the mother; she's just standing in for the "real" mother. The "real" one is the one whose body created the gamete that grew into that baby whose head emerges covered in the blood and fluid of the not-real mother. What is the reality and what the abstraction? Blood becomes an abstraction; seed remains reality.

And that is what I mean by saying that a patriarchal ideology dominates genetic thinking. The essence of that baby, who that baby really is and whom it really descends from is defined in terms of seed, even if the seed of women is also recognized in this new and modified patriarchal thinking. In this genetic model, women's contribution is reduced to the equivalent of men's contribution—the fact of the seed. To "mother a child" has been redefined and narrowed to become equal to "father a child."

Mapping the Genome

I write this just about one year since the mapping of the human genome was completed and announced with much fanfare. A scientific project that brought labs from around the world together, using the services of both public and private agencies, succeeded in its mission. Probably the last scientific accomplishment of this magnitude and concerted effort was putting a man on the moon. The atom bomb was probably the one before that. Let us take the man-on-the-moon as a point of comparison. It is, without a doubt, a pretty impressive accomplishment, just the kind of thing people never did think would happen in their lifetime. And yet it did: a man walked on the moon, planted a flag and, well, left. When you come down to it, what was there to stay for? It's a big empty rock in the sky.

Leonard Cohen put it in a song: "No, they'll never, they'll never reach the moon now. At least not the one they were after." Shooting for the moon, reaching the moon, touching and walking on the moon—a fabulous concept. Actually walking around up there in a space suit: ho hum. Like most of our tourist photos, it's the picture of us that turns out to be most fascinating: that photo of the earth taken from space, now *that* resonates.

Mapping the human genome turns out to have some of the same kind of "postpartum blues" quality. So what have we got now? Was it worth it? We have a long string of base pairs—"GAG A CAT"—the human genome, letter by letter. Most of it, the geneticists tell us, is "junk." Most of it seems to be the in-between stuff, with just nuggets of "genes," those magnificent, purposeful, creative, book-of-life, mysterious, unraveled bits. Early on, they—the geneticists doing this unraveling—told us there were going to be 100,000 such genes: a suitably majestic number. Apparently there are more like 30,000, and so it seems unlikely that there really is a gene for losing car keys, a gene for divorce, a gene for liking green apples. Thirty thousand just isn't enough to go around; a human has barely more genes than a worm has. We're going to have to explain it all with just 30,000 genes, and it's going to be a tight fit.

Maybe we should turn our attention to proteins? The new frontier is proteins: what genes produce and how those productions interact.

But we've got the genes. We've got them all spread out now, untwisting the DNA from inside the nucleus and spreading it out on page after page after page, available for your viewing pleasure online. And where does that get us? Is this the equivalent of another empty rock in the sky? Or is it indeed the way to solve human problems?

The Fabulous New World of Baby Making

When I first started following the media presentations of genetic science, much of the fanfare was under the general headline: "New Hope for the Childless." Understanding genetics, conception, embryonic development, and the ways that genes combined and played out was going to bring us a fabulous new world of baby making. People who simply could not get their gametes to turn into babies would be assisted. "Artificial reproduction" was the earliest language: cleverly, "assisted reproduction" was substituted, and an alphabet soup of baby making (IVF, GIFT, ZIFT and more) came into being.

Not only those who were "infertile" could be helped. Families who carried a genetic disease—like a curse, lurking in the genes, springing forth in generation after generation—could also be helped. New and healthy babies would be brought into the world; damaged, hurt, cursed babies could be prevented. "Prevention" turned out to be a bit complicated. In this context, it most often became synonymous with abortion: conception could not be prevented, but the products of conception could be tested, and those that were not acceptable could be eliminated. My fingers hesitate over the word "eliminated," as well they might, in this tense American climate regarding abortion. A newly fertilized blastocyst on a petri dish or a fetus at twenty-two-weeks in a normally occurring pregnancy post-amnio: each of these could be prevented from developing further—"eliminated." Ended? Finished? Find me a neutral word.

The experience of prenatal diagnosis and selective abortion is itself not a neutral experience. It causes individual women enormous pain and suffering. While "medically indicated" abortions are among the very most socially acceptable, far more so than having an abortion simply because one chooses not to be pregnant, they are not the most psychologically acceptable. A usual abortion is about getting "unpregnant." The woman wants simply to return to her nonpregnant status. The abortion that follows prenatal testing requires not that a woman avoid motherhood, but in fact that she embrace it. The decision making, choosing to end the pregnancy because of what the potential child would face, requires a distinctively maternal stance. It is precisely as the mother that she chooses what to do for her child-to-be or child not-to-be.

And yet, like mothers everywhere, women do not choose the circumstances in which they will raise their children, or the circumstances in which the child will live its life. Some of the conditions being tested for are simply unthinkably awful: they cause more physical pain for the baby than a mother would ever ask her child to bear. But most do not: people with Down syndrome, the condition around which testing and selective abortion was largely developed, can lead quite pleasant, even happy and productive lives. With support. With services. With money.

Does a woman given a diagnosis of Down syndrome in her fetus, without the support, services or money with which to raise that child, have choices? Is she empowered under these circumstances?

Women are offered prenatal diagnosis and selective abortion in the context and language of "choice." Genetic counselors take great pride in being "nondirective," which has become their mantra. The women I have interviewed over the years did not experience prenatal diagnosis and selective abortion as a situation of choice. (Rothman, 1993). They reacted to a series of constraints. Women in that position, whatever choice they made—to test or not to test, to abort or not to abort—often resorted to a phrase that continues to haunt me. Whichever choice she made was "my only choice." An only choice: a decision one reaches when there are no real choices, in a situation where the language of choice is the only language spoken. This is not a choice that empowers women.

One path that the new genetics led us on was thus straight to abortion, not a pretty place to go in America—certainly not the place to go if you want lots of funding to engage in a huge, multiyear, multinational, partially publicly funded research project.

The other path, the path of the new reproductive technologies, also became complicated. Certainly poor Mary, or Susan, got to have a miracle baby in her arms. Married safely to the father of the baby, with lots of money to spend on getting pregnant in the first place, and lots of money to lavish on the child—or the two or three or six children that the technologies sometimes produced—Mary or Susan (and never both, not the egg donor and the surrogate, just the one that got to be the new Mom) graced the covers of Sunday supplements the world 'round.

But one such Sunday supplement (Murphy, 2002) carried a different kind of miracle baby story: a deaf lesbian couple approached their sperm bank asking for a deaf donor. Congenital deafness, they were told, was precisely the kind of thing sperm banks screen out, not in. They approached a deaf friend instead. All they were trying to do, they reassured the reporter and the readers, was increase the odds of having a deaf baby: they were entirely prepared to love their baby no matter what its hearing status. But they would prefer a baby who is deaf, a baby like them, a baby who will grow up in their community and culture. As, after all, who wouldn't?

That most of the focus of the story and of the response has been on the deafness issue, not on the fact that they were a lesbian couple, probably represents some kind of progress. But lesbians, gay men, unmarried women of all sexual orientations, women whose husbands died and left sperm banked behind them, couples who split and left embryos in the back of some freezer somewhere—they all began to ask for "assistance" in their reproduction. The stories of new reproductive technologies took this turn. No longer just the domain of happily-married-but-desperate-for-children couples, the new reproductive technologies attract consumers who are creating families that, shall we say, pose a "challenge" for those pushing family values. The intent was to take care of the desperation of the childless, empower those (married) would-be mothers. But this? This is getting out of hand.

And so I have noticed over the past few years that reproductive genetics is no longer what dominates the headlines when the genome project is being lauded, celebrated, pushed. Now it is "pharmacogenetics," using the new genetics to cure disease. We thought we were all "pro-motherhood," that nothing could be more American and apple pie than dear old Mom. But now that dear old Mom might be a deaf lesbian, a gay man hiring a surrogate, a woman in her sixties, or who knows who, empowering her is a far different thing, and we need to find something else to agree on. Cancer? Are we all opposed to cancer? Well, OK then, let's use the new genetics to fight cancer!

Cancer as (Not) a Genetic Disease

Cancer is the disease of the new millennium. It is not, oddly enough, AIDS, which is, in its own way, a very old-fashioned disease. AIDS is spread by contact with an infectious agent, much like polio (see Martin, 1994, for a history of Polio and AIDS). Polio was spread by the innocent play of our younger children; AIDS spreads by the more subversive play of our older children. But with AIDS, as with polio in its time, danger comes from outside, since safety is believed to lie in the monogamous, heterosexual, nuclear family, inside a carefully cleaned circle.

The "genetics" of HIV that we are learning is the takeover story. Like a computer virus, the infiltrator "takes over" the cells, rewrites the commands, and shuts down the system. The "genetics" here is not part of the larger morality play. It is just the mechanism the infiltrator uses. Polio, when you think about it, didn't eat away muscles. It, too, was a virus that rewrote cell instructions. But that's not the way people thought about it at the time. The story, the plot line—with AIDS as with polio—is that something comes in, attacks, and destroys.

It is cancer, and especially breast cancer, that is writing the new story, moving us forward into the new thinking, the new geneticism. Breast cancer is the competing disease, the *other* plague of our time. The competition with AIDS comes right to the fore in the ribbon display: in response to the angry, flaming red ribbon of AIDS, breast cancer offers us the pink ribbon. If AIDS is the disease of dirty boys and their innocent victims, breast cancer is the disease of innocence, of mothers, grandmothers, aunts, and sisters. It is a disease that grows at home, not out there in dirty places. It is the disease that both reflects and in turn reinforces the newer, genetic model of disease. Cancer is now understood as a disease not so much of organs as of cells, a disease of the program of the body, a genetic disease.

Ordinarily, when people speak of something as "genetic," they are speaking of it as inherited, as part of the original program that turns the fertilized egg into the grown person. When geneticists speak about genes, they are speaking of stretches or segments of DNA. Those segments, when in the sperm and the egg, are indeed our "inherited genes." Those segments in a breast cell, a prostate cell, or a lung cell are still genes to a geneticist, but not "genes" in that sense of history moved along from parent to child. When geneticists study cancer, they are observing, studying, trying to understand how the DNA segments operate in the cancerous cells. Very occasionally, this has to do with the DNA segments as they were "passed on," in the sense of inheriting a "gene for" cancer. But mostly what geneticists look at are the mutations, the changes, that turn cells of individual people from normal cells of the body—ordinary breast, prostate, or lung cells—into cancerous cells spreading within the organ and metastasizing throughout the body.

From the perspective of the geneticist, it makes perfect sense to think of cancer as a genetic disease: it is a disease that occurs within the DNA of the cells, and an understanding of the changes within the DNA segments will perhaps provide an understanding of the cancer. But that is *not* the same as thinking of cancer as a "genetic disease" in the old sense that preceded modern genetics—as an inherited, passed-along-in-the-family disease. Very few cancers show signs of being "inherited" or "genetic" in that sense.

And yet, I find that that is not the way the story is being heard. A number of scholars have been observing how genetics is presented in the media (see, for example, the work of Peter Conrad). The frame for genetic stories is the finding of "the gene for": that is the plot line. Then there are some holes to fill in, some finer stuff about what percentage or what variation of the disease, how common the gene is, and how useful this will or will not be for prevention or for cure. But the basic story line is that they are finding the genes that cause disease, and cancer is a genetic disease.

Much has been made in the media of the "breast cancer genes," BRCA1 on chromosome 17 and BRCA2 on chromosome 13. Together, these genes account for perhaps 5 percent of all breast cancers and seem to account for 90 percent of all of the inherited cases of breast cancer, all the breast cancers that "run in families." Other genes, not yet found, may account for the other 10 percent of these inherited breast cancers, but all the inherited breast cancers together still leave about 95 percent of breast cancers unexplained.

And yet people seem to have come to believe that breast cancer is a genetic disease, a family disease, and that some of us are doomed. Women who have had one grandmother develop breast cancer in her old age tell me they are "at risk" for breast cancer because it "runs in the family." A seventy-year-old woman, an aunt of one of my graduate students, was told by her family doctor to have both of her healthy breasts removed because her sister developed breast cancer. If it weren't so tragic, if people weren't so terrified, we would say this was getting silly.

Breast cancer is a disease of largely unknown origin: the overwhelming majority of breast cancers are idiopathic, seemingly random, choosing this woman rather than that one not because of some family curse but perhaps because of where in the world she lives, perhaps the foods she has eaten, perhaps how those foods were grown. The truth is, we simply do not know what causes most breast cancers.

And what of the women who *do* carry this gene? What does the technology offer them? Well, now, that's not certain. They could be tested more frequently, but we don't know if that will really help. They could have their breasts removed, but even that might not entirely solve the problem. Removing healthy breasts when there are better-than-even odds that the woman would never develop breast cancer is a pretty drastic version of "preventative medicine."

In short, are women empowered by this approach to breast cancer? I do not think so. Are genetics laboratories, pharmaceutical companies, scientists, researchers, and the entire industry built up around genetics empowered by linking their work to breast cancer? Why, yes, I do think so.

Conclusion

Can genetic technologies be used to empower women and to improve the lives of women? Certainly, at the individual level, the answer has to be yes, in some circumstances. A woman who is unable to conceive and very much wants to experience a pregnancy can obtain great pleasure from the use of a donated egg. And that egg might, just might, actually be donated, though more usually it is purchased from a young woman who takes considerable risks with her health to make that money. But, yes, matching a willing donor or seller and an eager recipient can result in a good thing happening.

A woman who has watched a sibling suffer from a genetic disease can take great comfort in entering a pregnancy with the knowledge that if the baby she carries has that same disease, she can terminate the pregnancy.

A woman who has watched close female relatives suffer and die with breast cancer might well want to be tested, to know if she risks that fate or if her risk is that of anyone else in the population. Maybe she will have a better chance of treating the disease if she develops it, and maybe not; maybe she will just want the information and hope that she has the wisdom to deal well with it.

But is this technology fundamentally in the best interests of women? I think not. Do we want to control our fertility? Certainly. What would help us do that? On a very long list, we might well want to include some of the new reproductive technologies. But access to basic health services, good nutrition, the kinds of childcare and career services that would make childbearing during the years of peak fertility more feasible, sex education, safer contraception—all these things would be higher priority on the list.

Do we want our children to be healthy? There is no question that we each want what is best for our children. And how can we make that happen? There are, of course, many ways. I do not have the numbers, and I am not going to donate a day of my life to this project, but consider how frequently little children, low to the ground as they are, disproportionately become the victims of cars by being run over. We know these accidents occur right in their very neighborhoods, even their own driveways. My husband once ran over a tricycle in our driveway, and I will never forget the sight of that empty mangled little trike—and what it might have meant.

But if I suggest that perhaps we should separate our residential space from cars by providing centralized parking a few streets away, people think I am being foolish, utopian, unrealistic. Yet we find it sensible to search for a gene that causes (less) damage and (fewer) deaths of little children than do cars. Of course we want to prevent the suffering of little children. But rearrange the parking? Get real!

Does an individual woman benefit from this? Sure—if her child would have had one of the bad genes they find. Do our children as a whole face less suffering? Well, no, not really.

Does a woman who carries BRCA1 or BRCA2 have some choices? Sure. She can amputate both breasts—that sometimes (not always) works. She can identify and abort female fetuses who would carry the gene themselves. Will this reduce the danger of breast cancer? Not for most of us, for whom the 1-in-9 figure is the low lifetime risk.

So the search for genes is not, I argue, part of a larger project we are engaged in to reduce the causes of human suffering or to empower women. Human suffering—especially of the photogenic kind—is a justification needed for supporting genetic research. The new genetics—the laboratories, scientists, and industries—need women as consumers and as patients more than we as women need the new genetics.

References

Conrad, Peter. (1997). Public eyes and private genes. *Social Problems* 44(2): 139–154.

Keller, Evelyn Fox. (1995). *Refiguring life: Metaphors of twentieth-century biology.* New York: Columbia University Press.

Martin, Emily. (1994). *Flexible bodies: The role of immunity in American culture from the days of polio to the age of AIDS.* Boston: Beacon Press.

Murphy, Liza. (2002, March 31). A world of their own. *The Washington Post* (n. p.).

Rothman, Barbara Katz. (1993). *The tentative pregnancy: How amniocentesis changes the experience of motherhood.* New York: W.W. Norton.

7

Some Unintended Consequences of New Reproductive and Information Technologies on the Experience of Pregnancy Loss

LINDA L. LAYNE

In the last decades of the twentieth century, new reproductive and information technologies changed the experience of miscarriage, stillbirth, and early infant death for many middle class American women, often in unintended ways. This chapter looks at how these technologies affected the incidence and etiology of pregnancy loss; expectations about pregnancy outcomes; prenatal bonding and the social construction of fetal personhood; the diagnosis and medical management of pregnancy loss; and delivery of social support to women after a loss. It concludes by diagnosing an acute case of technological somnambulism and traces the etiology of this problem to the patriarchal ideology that continues to infuse childbearing in the United States.

Methods

This study is based on ethnographic field research from 1986 through 1989 with three pregnancy loss support organizations: UNITE (a grief support organization established in 1974–75 and serving approximately 1,000 families a year in Pennsylvania and New Jersey), SHARE (Source of Help for Airing and Resolving Experience, established in 1977 in Springfield, Illinois, and today is the largest pregnancy loss support organization, acting as an umbrella for approximately ninety-seven support groups throughout the

United States), and the New York Section of the National Council of Jewish Women's (NCJW) support group in New York City (established in 1982–83). It is also based on a textual analysis of the SHARE and UNITE newsletters (from the early 1980s to 2003), a review of popular and social scientific literature on the subject, and the personal experience of seven miscarriages (1986–1995).

Changing Incidence and Etiology

The 1980s and 1990s saw changes in the incidence and etiology of pregnancy loss, many of which were technologically driven. In the United States, there was an overall increase in the number of pregnancy losses (up from an estimated 890,000 spontaneous fetal losses in the U.S. in 1992 [Ventura, Taffel, Mosher, Wilson, & Hershaw, 1995, p. 12], to approximately 983,000 in 1997 [Ventura, Mosher, Curtin, & Abma, 2001, p.1]). Some types of loss increased while others decreased. Several of these changes were due to new demographic patterns such as a later age for childbearing. The rate of childbearing increased for women over age 30 and declined for women under age 30 (Ventura et al., 2001, p. 1), and the rate of pregnancy loss generally increases as women age (Cunningham et al., 2001, p. 856).[1]

Other changes were the result of new reproductive technologies, which had, in many cases, developed in response to the reproductive consequences of an older childbearing population. For example, there was a dramatic increase in the number and rate of ectopic pregnancies (Cunningham et al., 2001, p. 885). The rate of ectopic pregnancy "increased fourfold from 1970 to 1992," bringing the rate up to almost 2 percent of all pregnancies by 1992. This increase is attributed in part to the use of assistive reproductive technologies such as in vitro fertilization (IVF), gamete intrafallopian transfer (GIFT), and zygote intrafallopian transfer (ZIFT), technologies that were developed in the 1970s and 1980s in response to demand from the baby-boom generation, who had postponed childbearing and then were confronted with the effects of age on their fecundity.[2] An earlier reproductive innovation, the administration of diethylstilbestrol (DES) to over three million women in the United States between 1938 and 1975 as a preventative for miscarriage, also accounts in part for this increase in ectopic pregnancies. DES daughters have "between 8.6 and 13.5 times more ectopic pregnancies than normal" (Dumit & Sensiper, 1998, p. 218). They also have a higher risk of miscarriage and "between 4.7–9.6 times more premature births than normal" (Dumit & Sensiper 1998, p. 218).

At the same time, there has been a decrease in the number of late losses (Tangir, 2001) due to the advent of neonatal intensive care and the increasing use of prenatal diagnosis. Neonatalogy became a pediatric subspecialty in 1975 (Budetti, Barrand, McManus, & Heinen, 1981, p. 8) and neonatal intensive care increased the survival rate for very low birth weight babies dramatically during its first decade. An Office of Technology Assessment report concluded that if in 1985, "neonatal intensive care were provided for all very low birth weight infants," more than 17,200 children who would have died in 1975 would survive. Of these, it was estimated that about 2,200 would be seriously handicapped (OTA, 1987). During the 1970s and 1980s, a number of forms of prenatal diagnosis (alpha-fetoprotein tests, chorionic villus sampling, amniocentesis, and triple serum) were introduced and routinized. A direct result of these technologies has been a decrease in the number of late-term losses. "The fetal death rate due to lethal anomalies declined by almost half" because of early terminations for these pregnancies (Cunningham et al., 2001, p. 1074).

However, some types of perinatal death have become more common as a result of new reproductive technologies. For example, the frequency of multiple gestations has significantly increased because of the use of fertility drugs and also with a later average age of childbearing (women over age 34 are more likely to spontaneously conceive multiples). Research indicates that about "10 percent of all perinatal deaths are multiple birth babies" (Launslager, 1994, p. 120).

The last quarter of the twentieth century also saw an increase in the number of pregnancy losses attributed to environmental toxins. For instance, in the mid 1970s, the elevated rates of miscarriage found in Love Canal, New York, and Yellow Creek, Kentucky (Couto, 1986), and stillbirth found in Woburn, Massachusetts, were correlated with exposure to toxic dumps; the elevated incidence of miscarriage in the Alsea region of Oregon was correlated with exposure to herbicides being used by private timber companies and the U.S. government (Layne, 2000). In the mid-1980s, the elevated rate of miscarriage documented in the "Cancer Alley" region of Louisiana was attributed to exposure from the chemical plants that line that section of the Mississippi River (Miller, 1994).

Shaping Expectations

Ironically, during these decades when the overall incidence of pregnancy loss was increasing, awareness of these events decreased. Demographic changes played a role here, too. During the time when women were pregnant dur-ing most of their childbearing years, every family was likely to experience multiple losses. It is no longer common, however, to hear about middle-class women like Kim Morrison's Irish-born mother who "conceived 20 offspring, seven of them either stillborn or dead within their first year" (Nixon, 2003).[3] Growing up in such large families inevitably exposed children to the realities of such losses. A gripping example of this was provided to me by Kate M, an Irish-American healthcare professional in New York City. Kate's mother had nine pregnancies, three of which ended in miscarriage. Kate is the eldest, and her mother's first two miscarriages occurred when Kate was too young to remember them; the third, she recalls vividly. One day in 1960, during her sophomore year in high school, Kate uncustomarily decided to skip gym class and go home earlier than usual. She found her mother "white as a ghost," passed out in front of the sofa in their living room, hemorrhaging on the pale green carpet. Kate got through to her mother's doctor, whose phone number she found on the refrigerator, and he instructed her to call an ambulance and to try to find the fetus and put it in a jar. She found an egg-sized fetus in the blood clots around and under her mother's legs and sent this along with her mother in the ambulance, staying behind to care for her younger siblings when they returned from school.

Smaller family size not only means that women are much less likely to be exposed to such losses during their childhoods; it also means, in conjunc-tion with increased geographic mobility, that fewer women now enter their childbearing years living in close proximity to a dense network of female kin who may have shared their experiences of loss(es).

The frequency of pregnancy loss is also hidden in most of the popular lay guides to pregnancy and fetal development. How-to books of pregnancy, which can themselves be considered technologies of reproduction, typically take a woman step-by-step through a pregnancy from conception to birth without making clear that a pregnancy may end at any point along the way. Pregnancies are broken up into a series of stages (trimesters, weeks of gesta-tion) and presented as if one stage *inevitably* follows another. For instance, *Getting Ready for Child Birth: A Guide for Expectant Parents* (1986) devotes one of its 285 pages to pregnancy loss—but not until chapter 7, which is entitled "The Newborn." *The Rodale Book of Pregnancy and Birth* (1986) neither includes figures on the frequency of miscarriage nor explains the normal medical procedures for miscarriage. It does mention miscarriage and stillbirth in its flow charts at the beginning of the book (mostly to warn women who have had a previous miscarriage to be more careful this time), but discussion of these topics is placed incongruously in the section on the third trimester. *Our Bodies, Ourselves,* widely known as "the women's health bible," has been, and continues to be, organized in a similar fashion.

Fetal imaging technologies like the innovative and widely viewed *in utero* color photographs by Lennart Nilson have been used to tell "the story" of human reproduction. The Emmy Award–winning NOVA documentary, *The Miracle of Life* (1986), featuring Nilson's photographs, "follows the development of the single new cell into an embryo, then a fetus, until finally, a baby is born." Although it does stress obstacles to fertilization, particularly those that occur during "the perilous journey" of sperm, it suggests that a live birth is the inevitable outcome once fertilization has taken place.

This narrative structure is also powerfully portrayed in the prenatal development exhibit at Chicago's Museum of Science and Industry, where forty fetus specimens are arranged in jars along a wall in developmental sequence (Cole, 1993). These fetuses obviously did not result in a live birth! Because of abortion politics, the museum does takes pains to reassure viewers that, "to the best of their knowledge," these fetuses did not survive due to "natural causes or accidents" (http://www.msichicago.org, accessed May 20, 2003). In other words, these specimens each represent an involuntary pregnancy loss. Yet, this educational exhibit does not inform viewers about what it is that they are viewing—i.e., apparently normal embryos and fetuses which, for unknown causes, did not survive. Instead of educating guests about the reality of pregnancy loss, a common and important biological event, these miscarried or stillborn embryos and fetuses are ironically made to tell the story of successful, step-by-step development, "the journey we all made from a fertilized egg to a complete human being" (http://www.msichicago.org, accessed May 20, 2003).

Adjacent to this exhibit is the museum's popular baby chick exhibit where one can watch chicks hatch. Placards for this exhibit inform museum guests that "the embryo starts to develop after it has been placed in the incubator. . . . The fertile egg develops into a healthy baby chick in 21 days." Here again, the fact that some fertilized eggs will not develop at all, let alone into healthy baby chicks, is rhetorically negated by a positive, normative assertion of inevitably successful outcomes. Yet, as any woman who has suffered a pregnancy loss knows, it is not wise to "count your chickens until they are hatched" (Cote-Arsenault & Morrison-Beedy, 2001, p. 242).

Another factor contributing to a diminished understanding of the frequency of pregnancy loss has been the advent of a widely publicized "revolution" in reproductive medicine. Dorothy Nelkin has traced the vicissitudes of interest in science reporting and the popularization of science, noting that during eras like the immediate postwar years, when federal research funding was expanding, "interest in popularization . . . waned." But when "demands for research funds . . . outstrip the supply," scientists and their institutions seek more press coverage and invest more in public relations (Nelkin, 1995, pp. 124–133). As "medical procedures became increasingly costly during the 1980s, technical institutions expanded their public relations efforts. . . . Public relations in medical institutions play to a receptive press eager to publish stories on the most advanced . . . experiment, the latest . . . therapy, or the newest . . . technology" (Nelkin, 1995, p. 131). Nelkin notes, "Research on possible therapies for a devastating disease is always newsworthy and often publicized far too soon" (1995, p. 132). During the 1980s, infertility became newsworthy as it acquired disease status (Sandelowski, 1993, p. 7). Dramatic coverage, such as the story "Miracle Baby," which appeared in *Ms* magazine in 1989, or the cover story of the December 1993 issue of *Life* magazine, "Miracles of Birth: The Blessings of a Medical Revolution, Healthy Babies Who 10 Years Ago Would Never Have Been Born," became a staple of national science reporting. The article in *Ms* describes the survival of a baby born at 23.5 weeks' gestation. The subtitle reads: "A stunning victory for neonatal technology, raises as many questions as it answers"—including costs: about 1 million dollars for the six-month hospital stay, and substantial pain and suffering for the child. Nonetheless, the article concludes with an optimistic view of this child's future: "'Isabelle's major problem in the future,' one of her nurses jokes, 'will be that she won't be able to wear a bikini' because surgical scars cleave her torso" (Halpern, 1989, p. 64).

Stories like these clearly reinforce the view that reproductive technologies can indeed achieve miracles. The *Life* piece begins, "Not so long ago, nature was in control. A woman got pregnant, checked into the hospital, gave birth and went home a week later" (a view echoing the story told in pregnancy manuals, popular science exhibits, and documentaries). It continues, "If she couldn't conceive or carry to term, or if something was wrong with the fetus, she had few options. Since then, year by year, doctors have learned to manipulate the mechanics of pregnancy and birth. The first test-tube baby in the United States was born in 1981; now children conceived in laboratories are born every day. Illness in a fetus can be diagnosed and cured before birth." It goes on to present "the exclusive stories of people who needed miracles—and found them" (Dowling, 1993). Media hype like this surrounding advances in reproductive technologies contributes to unrealistic expectations regarding biomedicine's abilities to guarantee a live birth.

Jim Friedrich (1984) poignantly expresses the widespread belief that if there is a problem during a pregnancy, doctors will be able to fix it. Even after having gone through one miscarriage, he retained his faith in science's power and described himself as "a joyful, expectant father" during his wife's subsequent pregnancy.

> ₄Things were looking bright [until one day his wife] Val said there was no more movement, and she became very worried . . . I knew, because I am the eternal optimist, that there was nothing to worry about . . . [The next day the doctor] could not hear a heartbeat. . . . I met her at the hospital . . . The doppler was hooked up. Three nurses were checking Valerie and the machine. The electronic wizard would pick up occasional sounds, but nothing definite. . . . "The baby must have died." . . . Valerie knew but I didn't want to accept it. Not with modern technology.

Carl Jones (1987) describes similar feelings of disbelief: "Surely this cannot be serious. Babies don't die nowadays."

Although fathers appear to be especially prone to believing in the efficacy of modern medical technologies, mothers are not immune. For example, Lisa Casimer (1987) writes, "We knew instantly that we had created [her] and we were especially careful to ensure that she would be a healthy baby. Naturally, because of our prenatal care, we assumed that we would experience a normal pregnancy." Tami White (2000) tells of how, after two miscarriages, she and her husband decided to "seek a specialist. . . . After many tests and 7 inseminations, I became pregnant again. We believed everything would be okay because we were in the hands of a specialist. Not so . . ."

Donna Brunner (1992), a woman who had a c-section for her first child then lost a baby trying what is known as a VBAC (vaginal birth after c-section) writes, "In our day of advanced technology, we seem to believe knowledge will conquer all—even fear, pain, labor, contractions. But it didn't work for me either time."

Facilitating Prenatal Bonding

New reproductive technologies have also accelerated the pace of prenatal bonding. Because of technologies like home pregnancy tests and sonograms, many women and their social networks are beginning to construct the personhood of their embryo/fetus at earlier points in a pregnancy.[4]

The vast majority of pregnancy losses occur during the first weeks of a pregnancy (about 10 percent occur before the first missed menses). Whereas until the 1980s pregnancies were normally medically determined by a blood test conducted in a doctor's office often a month or more after the first missed menses, now most middle-class women use over-the-counter, do-it-yourself kits (Layne, 2005). These affordable, low-tech, easy-to-use tests measure the pregnancy hormone (human chorionic gonadotrophin, HCG) levels which begin to rise with conception, and so can establish a pregnancy even before the first missed menses. "Clear Blue" made by Unipath in the United Kingdom, and "First Response," made by Armkel in California, both boast "earliest results," "results four days sooner than other leading brands." (Clear Blue reports 53 percent of pregnant women tested positive four days before first missed menses; 74 percent three days, 84 percent two days, and 87 percent one day earlier). These tests mean that women are experiencing themselves as pregnant much earlier in a pregnancy than had been the case in the past. This is undoubtedly changing the experience of those who have early pregnancy losses.

Accounts of home pregnancy tests serve as the opener in many narratives of loss. For example, Jennifer Fisher (2002) begins her story by describing her feelings. "When I took a home pregnancy test on August 10, 2001, and found out I was pregnant. . . ." And Chrissy Coggins (2000) starts by describing the large role that these tests had come to play in her quest for a child: "For the last eleven years, I'd been trying to conceive. Month after month, prayer after prayer, and one home pregnancy test after another, it was all I could think about. While my peers were raising children and getting pregnant, I was crying. . . . At one point, I believe I psychologically made myself pregnant. I had all the typical symptoms, but I still got a minus sign." She then tells of how, one December, she finally decided to "accept the fact that I would never be a mother, but then on January 27, 1999, my dream came true . . . I saw a plus sign! Hallelujah! My first thought was 'Thank God I don't have to buy anymore of those pregnancy tests' (I should have been buying stock in them instead). My second thought was 'Is it really a plus sign or are my eyes playing a trick on me?'" She woke her "significant other" so that he could confirm the test result. "I had blood tests done to verify everything." Then they "began putting a plan together for their future."

Fetal imaging technologies, particularly obstetrical sonograms, have also been a major innovation in terms of prenatal bonding. The import of prenatal scanning in terms of bonding was recognized early. In 1983, Fletcher and Evans reported "two cases in which women in the late first or early second trimester of pregnancy reported feelings and thoughts clearly indicating a bond of loyalty toward the fetus that we and others had associated only with a later stage of fetal development" (p. 392).[5] They concluded that parents viewing their fetus via ultrasound "will experience a shock of recognition that the fetus belongs to them," that such experiences will result "in an earlier initiation of parental bonding" and are "likely to increase the value of the early fetus for parents who already strongly desire a child" (p. 392).

Since Fletcher and Evans' observation in the early 1980s, when obstetrical ultrasound was relatively rare (Taylor, 1998, p. 17), the use of sonograms during pregnancies has increased dramatically.[6] According to Taylor, "It is probably safe to say that in the United States, nearly every pregnant woman who has access to any form of healthcare will have at least one ultrasound scan during pregnancy" (p. 19).[7] Women deemed "high risk," a category which includes women who have had a previous loss or who are undergoing fertility treatments, have many more scans—up to one a week during the first trimester for the purpose of "reassuring" them, yet as noted earlier, these women are much more likely to experience pregnancy loss.

A number of anthropologists have noted the important role sonogram images play in helping to establish the "reality" of a "baby" during the early stages of pregnancy. Nearly all of the forty-nine Canadian women Lisa Mitchell (1994) interviewed during their first pregnancies "talked about the fetal image as a form of proof, saying 'Now I know I'm really pregnant!' 'Now I know its real!' or 'Now I know there's a baby in there.' Having this proof enabled women to talk more confidently about 'the baby' and avoid the term 'fetus' altogether" (p. 153).

Ultrasounds are often used by pregnant women to enlist others in the social construction of their "baby" (Rapp, 1997; Taylor, 1996, 2000a, 2000b; Mitchell, 1994). The bonding that is facilitated by these images seems especially important for others, especially fathers-to-be (but also future siblings, grandparents, etc.) who now have direct access to "their baby." Several men have told me how much they liked sonograms because it made them feel more a part of the pregnancy.

Prenatal bonding facilitated by sonography can take place even before a pregnancy occurs. Sonograms are used in conjunction with ovulation stimulation for a variety of infertility treatments (e.g., in preparation for in vitro fertilization, artificial insemination, or simply, as in my case, to try to assure that conception took place at the optimum time while the "eggs were fresh"). Would-be parents are routinely shown the growing follicles on the sonogram screen. Every day or two the would-be mother (and perhaps the would-be father, too) witnesses the gradual growth and development of her egg/s and anticipates the climax of ovulation when the mature egg will be "expelled."

Clearly, this process can be seen and experienced as analogous to pregnancy and birth. One woman describes how "It really drew me in when I saw those eggs on the surface of my ovary . . . ." (quoted in Lasker & Borg, 1987, p. 56). Ripening follicles may be counted by would-be parents as the first stage of a pregnancy and sometimes attributed potential or quasi-personhood. Each failed attempt at conception may be experienced as a pregnancy loss. As one psychologist observed, "Some women become attached to those embryos in a way that's very similar to how attached women get to a pregnancy . . . [and they experience] a failed cycle of in vitro as a miscarriage" (quoted in Lasker & Borg, 1987, p. 59).

One man describes the experience of attempting to conceive with the use of fertility treatments in this way: "Every month for three days it's like a funeral. We've had twenty-four funerals" (quoted in Fillion, 1994, p. 50). He went on to explain, that he did not share his wife's feelings of grief, "It was as though a close friend of hers, who I'd never met, had died. My feelings were primarily a reaction to hers" (p. 50).

Some also view, via microscope, embryos that have been fertilized in vitro, and at some clinics couples are presented with a Polaroid of their embryos before "transfer" (Treichler, Cartwright, & Penley, 1998, p. 17, n. 8). One man tells how, when he told his wife that he wasn't sure what he was seeing, she responded: "What's the matter—don't you even recognize your own kid?" (quoted in Lasker & Borg, 1987, p. 67). Becker (2000, p. 121) reports one couple undergoing IVF who said, "We had embryos and we were putting them back, and, you know, you want to name them as they go in." Another couple told her how they had gone on vacation after an embryo transfer while waiting to see if any had implanted. They named the embryos and sent postcards to family and friends saying that "the embryos seem to be enjoying it here" (Becker, 2000, p. 156).

A heartbeat can be seen easily on a sonogram screen as early as eight weeks' gestation (i.e., six weeks after conception). Even when other parts of the embryo are hard to recognize, the pulsating heartbeat stands out amidst the grey. Anna Cooley (2002) describes such an experience in a piece written to a miscarried "baby": "I saw you on that ultrasound screen and I know you were real, you were there. . . . You may not have looked much like a baby, but I knew you were there, I could see your heart beating so strong."

That the heartbeat, a visible and audible sign of life, has a powerful effect, is evident in narratives of loss. Cari Simons, writing a year after a miscarriage at fifteen weeks' gestation, remembers: "We saw you at ultrasound and heard your heart beat. You were absolutely perfect from your head to your feet!" (Simons, 1995). Lori Carlin, who was implanted with three, eight-cell embryos, remembers how, at the first ultrasound at 4.5 weeks gestation, they saw two sacks, but at 7.5 weeks, "the ultrasound showed three little babies and we saw your little hearts beating." At five months' gestation she delivered, and by the next day, all three were dead. In a piece addressed to her children she writes, "I'll always remember seeing you on ultrasound, hearing those strong heartbeats."

Later sonograms and amniocenteses often reveal the sex of the fetus, and this knowledge is critical in terms of the construction of fetal personhood (Taylor, 2000a; Cartwright, 1993).[8] Often, the "baby" is named at this point and shopping, another primary activity by which parenthood and personhood are constructed, typically accelerates once the gender is known. The nationally covered Laci Peterson case illustrates this (Layne, 2004), as do accounts like that of Anne-Marie Lillyman (2001) published in pregnancy loss support newsletters. Anne-Marie explains, "My husband and I lost our little William at 20 weeks, just one week after the ultrasound confirmed he was a boy." Jennifer Fisher provides another example. She tells how, after the results of her "triple screen test (for Down syndrome and spina bifida) came back a little high, they wanted me to have an in-depth ultrasound." After that she was advised to have an amniocentesis "to make sure nothing was wrong. . . . By the early part of November, I knew everything was fine and I was having a girl. My fiancé and I were so excited. We decided to name her after his grandmother and my grandmother, Catalina Pearl" (Fisher, 2002).

Dana and Travis Hicks's (2001) piece, "Our Story: In Memory of Lance and Chase Hicks," also illustrates the role of technology in the construction of fetal personhood. "Our story began in June 1995. That was when I found out that I was pregnant with our first child. Like all expectant parents, we were immediately looking forward to our baby's arrival . . . . We were already making guesses as to whether I would have a boy or girl. We started trying to come up with names that we both liked and how we would decorate the nursery. . . . At twenty weeks we had an extensive sonogram. Our baby looked perfect and we learned that I was carrying a boy. We were very excited and decided on the name Lance Zachary." He died *in utero* several weeks later from a cord accident. Hicks was pregnant again in two months. "Again at 20 weeks I had a sonogram and learned that I was again carrying a boy. We decided to name him Chase Lucas." At twenty-two weeks she went into premature labor and had a stillbirth. During her third pregnancy she had a cerclage [a suture around the cervix to hold it closed] at sixteen weeks in an effort to prevent premature labor. At twenty weeks another sonogram suggested "everything was perfect .... This time we didn't want to know what I was having."

These technologies, precisely because they encourage prenatal bonding, often provoke strong negative feelings or deep ambivalence in women during subsequent pregnancies. Heather Gail Evans-Smith (2002) describes the feelings that the home pregnancy test evoked, "The test is positive and yet I am not overjoyed, But in pain, in fear of all the what ifs/The memories of before burrow through my brain/I sink to the bathroom floor/EPT test still in my hand/It is positive I whisper to the walls/It is true I am with child . . . I shake my head in despair/. . . I am pregnant and don't know what to say/I don't know what to think/I don't know what to feel/. . . As I lay on the cold tile/ Grasp the test in my hand /I wonder if this time . . . I will hold a baby in my arms" (Evans-Smith 2002). Jennifer Fisher (2002) explains her mixed feelings upon discovering with a home pregnancy test that she was pregnant again following a miscarriage at six weeks' gestation the previous year, "I was excited but also scared. I was scared that I was going to lose this one too."

The same holds true for sonograms during a subsequent pregnancy. Although sonograms are of little medical utility, their routinization has been rationalized by doctors because of the psychological benefit of "assurance" they purportedly provide pregnant women (Taylor, 1998). Taylor (1998, p. 19) found that "obstetricians and sonographers alike make frequent reference to ultrasound's 'psychological benefits,' both in informal conversation and in medical literature." These benefits, Taylor observes, paradoxically include promoting maternal bonding and unconditional maternal love while at the same time offering "reassurance" that the baby is "normal," which ironically speaks to the fact that it might not be and if it isn't will probably be aborted (p. 20).

During a subsequent pregnancy, the possibility of reproductive disaster is ever-present in women's minds (Layne 2006). These women are subject to more technological surveillance and less likely to be reassured by this technology. One of Cote-Arsenault and Morrison-Beedy's (2001, p. 243) respondents tells of how, when her doctor "was making me have a repeat sonogram and then said 'Don't worry,' . . . [she sarcastically replied,] "[Yeah.] Right!"

The extent to which women both need to and are unable to be reassured appears to be related to the amount of personhood that the would-be mother had assigned to the fetus during the pregnancy that ended in loss (which in turn is related to, but not synonymous with, length of gestation) (Cote-Arsenault & Dombeck, 2001). Cote-Arsenault and Dombeck (2001) found that the higher degree of personhood that had been assigned during a pregnancy that ended in perinatal loss, the higher pregnancy anxiety there was in the subsequent pregnancy (p. 660).

One strategy for dealing with the keen understanding of the provisonality of each pregnancy is to consciously make an effort to cherish each moment of the pregnancy. Cote-Arsenault and Morrison-Beedy (2001, p. 241) found that "some women wanted to enjoy the moment because the future was in doubt." For such women, technologically mediated access to their child via sonograms or fetal monitors may be welcomed as a way of augmenting their experience of the child. An example of this is found with Kris Ingle, a practicing nurse who became a leader of UNITE following her stillbirth. She describes her feelings during the ten-weeks' sonogram in her subsequent pregnancy (1986/87a): "I pause to cherish this moment. Today I saw your heart beat and your feet kick as your body curled and uncurled. The doctor tried to measure you, and you didn't cooperate. Then he was finally able to do so. He called you 'my baby,' as we watched your acrobatics on the ultrasound screen. . . . I'm thankful each hour that my body gets to cradle you."

Cote-Arsenault's studies of subsequent pregnancy support my own observation that more often women try to protect themselves "from the emotional pain of another significant loss by not assigning personhood" or at least not as much personhood to a subsequent fetus (Cote-Arsenault & Dombeck, 2001, p. 661). One of the strategies women adopt is to avoid exposing themselves to those aspects of prenatal technologies that enhance attachment to the pregnancy. They may opt out of learning the sex of a child and/or may try not to look at the sonogram screen.

Kelly Gonzalez (1988), coordinator of the SHARE group in Colorado Springs, describes how, during the pregnancy that followed the loss of her first child at twenty-two weeks, she tried, though ultimately with little success, to refrain from bonding during the pregnancy by avoiding technologically mediated access to her baby. She described how she felt during a ten-week prenatal checkup:

> I found myself not wanting to hear this new heartbeat. I wanted to shut it out. As the doctor put the stethoscope to my ears, I quickly took it off, after listening for only a few seconds. I realized I was trying to stop myself from getting attached to this new life inside of me . . . . At 12 weeks I had an ultrasound. Once again, I was filled with memories of the last time I was hooked up to an ultrasound machine, when the doctors ultimately learned that Alycia was dead. As the technician started the procedure, I avoided the screen. I did

not want to see my little baby moving and kicking, full of life. Curiosity got the better of me and I eventually watched the little gymnast at play. But the moment I realized that I was smiling and laughing, I quickly turned away, trying to stop myself from any attachment. (Gonzalez 1988)

Others explained how they "worried day-to-day about the well-being of their unborn babies" during subsequent pregnancies and how little check ups reassured them. One woman speculated that if she could have had round-the-clock technological access to her baby's heartbeat, i.e., if "they would have given me one of those things (Doptone) to take home so I could just hear the baby's heartbeat, I would have slept. But I didn't sleep" (Cote-Arsenault and Morrison-Beedy, 2001, p. 242). In fact, one company, BellyBeats, now offers Fetal Doppler-Baby Heart Monitor rentals. They describe their product as "a sensitive instrument that can detect the baby's heart rate as early as eight to twelve weeks into your pregnancy, providing you peace of mind that your baby is safe—well before your first doctor's appointment and throughout your pregnancy" (http:/www.bellybeats.com).

Some seek more frequent medical testing, not to reassure themselves, but to prepare for the next anticipated loss. For instance, Cote-Arsenault and Marshall (2000) tell of one woman who demanded repeated HCG tests during her subsequent pregnancy. After a positive home pregnancy test she "went in and had the blood test. I wanted an HCG then. I didn't want to do the pee in the cup. I wanted an HCG then. I wanted an HCG in a week. You know, it was like I was going to track this one and find out if there is any demise coming" (p. 482).

Barbara Katz Rothman (1993) has described how the waiting period for amniocentesis results produces "a tentative pregnancy." She explains that because of the strong likelihood that women will end their pregnancy if the results come back "positive," women wait months "in limbo . . . unsure whether they are 'mothers' or 'carriers of a defective fetus'" (p. 7). For women who have had pregnancy losses, this tentativeness does not end with good prenatal test results, however, but dogs a pregnancy until a successful conclusion (i.e., if they are lucky, for nine long months). As one woman described her pregnancies after miscarriage, "Worry is sort of a constant ache. It's just . . . eating at you all the time" (Cote-Arsenault & Morrison-Beedy, 2001, p. 242). (In my case, this tentativeness and wish not to become too attached so as to prepare myself for yet another devastating loss persisted even after Jasper's birth, during the two months he slipped in and out of critical condition in the neonatal intensive care unit [Layne, 1996]).[9]

Diagnosis of Pregnancy Loss

Whereas in the past a woman would learn that she was "losing the baby" by physiological changes in her body (bleeding and cramping, premature labor, the absence of kicking), now it is frequently through devices such as electronic dopplers or sonograms (or both) that would-be parents learn that their wished-for child has died (or, in fact, never lived). Pregnancy tests, whether urine or blood, only establish a chemical pregnancy. A sonogram in the first trimester determines whether there is a physiological pregnancy as well as a chemical one—i.e., whether there is actually an embryo or only hormonal changes that usually, but not always, signal a pregnancy. A first trimester sonogram may reveal an embryo that is alive (has a heartbeat), an embryo that is dead (does not have a heartbeat), an empty embryonic sack, or no sack at all.

Many women have experiences similar to mine. At one prenatal visit they see and/or hear a heartbeat. At the next visit, where there had been a magical tiny flicker of life on the screen, the screen is deadly still; where the room had been filled with the galloping-horse sound of the fast-paced fetal heartbeat, there is thundering silence. As Elizabeth Cohen described in her piece "The Ghost Baby," published in 1997 in the *New York Times Magazine,* "no one can tell you what it feel like to hear silence where the fetal heartbeat is supposed to be." In Cohen's case, she and her husband had "gazed in awe at our baby on the sonogram . . . . We heard his heartbeat, and then he put a thumb into his mouth and gave a playful kick of his leg." But just a few weeks later, when at her twenty-week checkup the midwife could not find a heartbeat on the fetal monitor, a second sonogram confirmed that her "son had died inside" for "no obvious reason" (Cohen, 1997).

For others the experience is more protracted. For example, after naming their daughter Catalina Pearl, Jennifer Fisher (2002) had a follow up "in-depth ultrasound" in her fourth or fifth month because at the first one "Catie was not big enough to see everything the doctors wanted to see." During the second ultrasound "we found that there was something wrong with Catie. She wasn't growing like she was supposed to. There was something wrong with the placenta, but we didn't know what the problem was, so the doctor wanted another ultrasound." A month later, during a third ultrasound, they "got the most devastating news. Catie had only gained three ounces, there was reversed blood flow in her umbilical cord, and there was very little amniotic fluid." Jennifer was told that whether she delivered immediately or waited, her child had virtually no chance for survival. "I made the decision to let my little girl live the rest of her life in peace. On January 8, 2002, I went to the doctor to make sure little Catie's heart was still beating. That was when we found out her heart had stopped." Labor was induced that night and her daughter born at twenty-six weeks, five days.

Recall Chrissy Coggins (2002), who, after eleven years of trying, finally got a positive result on her home pregnancy test. She ends her piece explaining that "Two ultrasounds, three blood tests and eight weeks later I was told 'there is no heartbeat and your hormone levels are dropping'. . . . My fetus was dying."

Often losses are discovered at "routine" ultrasounds that families anticipate with pleasure. Suzanne Kiper (2002) writes, "I had looked forward to [that day] for so many weeks, and at last it had arrived. Now in my fourth month of pregnancy, my husband Tim and I were joyfully anticipating our baby's ultrasound and all the wonderful glimpses into that silent, secret life that such technology made possible. . . ." Instead of these pleasures, "the solemn expression on the face of the ultrasound technician" and "the doctor's somber news" that their daughter was not growing and had less than a five percent chance of survival . . . brought devastating grief." Weeks later, at twenty-four weeks' gestation "there was no heartbeat; the ultrasound monitor revealed her tiny form, still and sleeping."

Similarly, Rita Fadako (1997), whose pregnancy lasted "twenty weeks, four days" recalls bringing her husband and mother with her for the first sonogram "to have your first pictures taken. . . . They must have taken around a hundred pictures and that is when I knew something must be wrong. The doctor explained that the baby's brain had not and would not fully develop and that there was no hope." Another example is Anita Horning (1997), who tells of taking her three-and-a-half year old son to the sonogram appointment and telling him he would "see a picture of the baby in Mommy's tummy" but the doctor said "he could not find any 'fetal pull.' I did not understand what he meant. But then he said there was no heartbeat and our baby had died."

Ultrasounds often reveal multiple gestations very early in the pregnancy, only to reveal at subsequent ultrasounds that one or more has died. For example, Marie Keeling (1987) describes "overwhelming joy when the ultrasound, at six weeks, showed twins! That joy turned to fear. Three weeks later—one of the twins had miscarried." Similarly, Tara Niles (1996) writes, describing the death of her daughter, Alli Renee, who died in January 1995 and was born June 6, 1995, along with her surviving twin, "I looked at your motionless heart on the screen. This can't be happening, is it just a dream?" Since it is quite common for one or more embryo to be reabsorbed during a pregnancy, early detection of multiples via ultrasound means that would-be parents are confronted with the grief of pregnancy loss that, in the past, they would have been spared. This is similar to the way that early detection of pregnancy by home pregnancy tests means that more women experience a pregnancy loss, whereas in the past they would not have even known they were pregnant; but in the case of twins, would-be parents must cope with the grief while simultaneously hoping for a successful birth. Knowledge of a lost twin is also likely to have implications for surviving twins, who often grow up with the identity of a twin but *without* their twin and may be faced with survivor guilt and the pressure of living for both (Blizzard, in press).

In losses that occur later in a pregnancy, after a woman has felt fetal movement, the impact of some of these technologies is truncated by the woman's own experiential knowledge of the baby she is carrying. In these cases, the technologies may confirm what a woman already fears: "When I realized I hadn't felt the baby kick for over 24 hours, I got really frightened. The next day I went to the doctor. He found no fetal heartbeat. I cried. . . . the diagnosis was clear—the baby had died *in utero*" (Iacono, 1982).

However, as Julie Gainer (2000) so vividly expresses, even later in pregnancy, the absence of movement for a time can be rationalized, and the ultrasound brings shocking news. "Then you grew so quiet and your joyous dancing slowed./ I convinced myself you were just sleeping, still warm and safe and huddled./ My strong and healthy Aidan James, just resting peacefully./ Then your tiny heart screamed its silence across every medical frequency."

It is also becoming increasingly common for ultrasounds to reveal a congenital defect that leads to death after a live birth and heroic medical interventions. Numerous accounts appear in the newsletters that are similar to that of David Weber (2001). David describes the death of his first child "who was born with a heart defect (hypoplastic right ventricle), which we had known about since the 'routine' ultrasound 17 weeks earlier." She died in infancy during her third heart surgery.

Medical Management of Pregnancy Loss

The technologically derived diagnosis of imminent abortions affects the mechanics of pregnancy loss. During the 1980s and 1990s, if a first-trimester demise had been diagnosed, the standard treatment was not to wait for the woman to miscarry on her own but to perform a dilation and curettage (D&C), the same procedure used for elective abortions.

One doctor explained his procedure for a patient who is miscarrying: "Get an ultrasound and see what you are dealing with. Find out whether there is a pregnancy there or not.... If you have spotting and you get an ultrasound and you find out that there is no longer a sac and essentially you already have dead products of conception, then there is no sense in sitting around in bed. ... do a D&C" (quoted in Pizer & Palinski, 1980, p. 111).

This doctor concluded, "You can show that there is a drop in chorionic gonadotropins to convince the woman that she lost the pregnancy." Today, even though a hormone tests would probably also be done, it is most often the static image of the sonogram screen that is used to convince a woman that all hope is lost.

Despite the sustained critique by the women's health movement of unnecessary obstetrical surgery (e.g., episiotomies and c-sections, not to mention gynecological surgeries like ovariotomies, cliterodectomies, and full mastectomies), feminists have not objected to this routine and almost always unnecessary surgery on women. It is only during the past five years or so that women are being given the choice of surgery, natural miscarriage (known as expectant management), or drug-induced miscarriage (known as medicated spontaneous abortion). (See Layne, in press, for a review of the medical literature on these options.)

The growing popularity of "expectant management" is in some sense an unanticipated consequence of the establishment and spread of infertility clinics. Endocrinologists who have clinics with on-site labs to check women's hormone have found another use for these technologies. After a demise, progesterone and HCG (human chorionic gonadotropins) levels drop gradually until the woman's cervix dilates and the uterus expels its contents. If any "products of conception" remain in the uterus, they continue to produce pregnancy hormones. Thus, by checking these levels doctors can be assured that all tissue was expelled.

Early losses are also now sometimes "medically managed." This technique involves giving women one of the two drugs used for the newly legalized medical abortions, a prostaglandin analogue (usually isoprostol) to stimulate the uterus to contract and empty. Women have follow-up visits with a clinician to be sure the abortion is complete.

I have described elsewhere (1992) other dimensions of the ways that the medicalization of pregnancy loss has changed the experience for women. In the past, one learned of the loss when one had a miscarriage or stillbirth. Today, diagnostic technologies cut up a loss into discrete moments: the moment of the demise, the moment of learning of the demise, the moment of birth or surgical removal. Now one can learn of the demise at a prenatal visit and schedule the surgical removal (via D&C) at the doctor's and/or patient's convenience This creates situations in which a woman may feel like "a human coffin" (Ingle, 1981/2) or "a living tomb" (Heil, 1982).

Documenting and Commemorating Loss

Sonogram images and the output from fetal monitors, along with post mortem snapshots, also serve as technologies of memory after a loss. Because pregnancy loss is a tabooed topic in our society, one of the most frequent complaints voiced at pregnancy loss support meetings is that relatives, friends, and co-workers do not understand the reality of their loss. For example, Barbara Cuce (1998), in describing the way people responded after the neonatal death of her son, asserts "It *did* happen and my baby was *real*, even though perhaps to them he wasn't." In the face of this cultural denial, parents often appropriate the authority of science and the objectivist power of photographic images to prove their claims. Sonogram images, scraps from fetal monitors, and snapshots are frequently saved by bereaved parents and utilized as evidence to prove to others that a "baby" existed. Kris Ingle (1986/87) reports "I told the nurse I needed a piece of each non-stress test reading, proof that the baby is there, heartbeat strong and beating, at least for today. . . ."

In 1987, SHARE conducted a "Baby Pictures" questionnaire and found that of 438 respondents who reported on losses that occurred between 1966–1987, 39 percent had ultrasound pictures; "93 percent of these said they were important keepsakes" (Laux, 1998a). Given the increased use of routine ultrasound during the past twenty years (see, for comparison, Rapp, 1997, pp. 31, 48), it is probable that the number of people with such photos is significantly greater today.

In addition to receiving sonogram images, women whose losses take place in the later half of their pregnancies or after birth are often presented with snapshots by the hospital staff.[10] A 1988 SHARE survey on the importance of photographs after a loss found that of the respondents 63 percent had one or more photographs, 50 percent were offered pictures by the hospital, and 26 percent had taken pictures themselves (Laux, 1988a).

Although at first these snapshots may not seem to qualify as new reproductive technologies, they clearly are parts of the new hospital-based system for medically managing pregnancy loss. Much as sonogram use is justified for the "psychological benefits" it provides parents (Taylor, 1998, p. 19), these photos are taken by the nursing or social work staff and offered therapeutically to the parents to aid them in the healing process.[11]

Like sonogram images, these photos play a critical role in establishing the reality of the baby. For example, one of the women in the self-help group featured in the film "Some Babies Die" explains that some of her relatives "took the view that it wasn't really a baby and they were not much support at all. If I'd had photos, I could have said, 'Look here, it was a baby. He was a beautiful child'" (Down, 1986). And a member of UNITE ends her poem: "One photo. One photo. Our special memory of your birth. A reality. A reality for those who tend to doubt your worth" (Burgan, 1988). The respondents to the SHARE survey also testify to the "reality making" function of such photos for themselves and for others: "We need to remember her as a real person we were holding. This is our proof"; "Although I get no great comfort from her pictures, I do have them put away for when I do feel the need to see she existed"; "On those days when you feel like it really never happened and people are treating you like you never had a child, you do have a picture to remind yourself you did have a child"; another says that a picture of her 20–week-gestation baby "hang[s] on the wall with the rest of the children . . . . It is proof she existed" (quoted in Laux, 1988b).

In addition to affirming the reality of a baby, such photographs are used by bereaved parents to stress the uniqueness and individuality of their baby while at the same time provide important resources for establishing family ties through the rhetoric of inheritance. They permit would-be parents to indulge in the postnatal American ritual of attributing family resemblances.[12] Parents who spend time with their child after her/his birth—whether the child is still alive or has died—engage in this practice. For example, Julie Caiola (2000) writes of her daughter, Angel, who lived an hour and twenty minutes after her birth at 21 weeks' gestation, "A mouth like your father's/But you had your mommy's nose."

Although the taking of such photographs has become standard practice (the American College of Obstetricians and Gynecologists recommends it in "Guidelines for Perinatal Care" (Marosi, 1999), and most pregnancy loss support group members value these photos), some people say they just make things worse. This is the subject of a 1999 article in the LA Times that reported on a handful of lawsuits in the United States challenging this innovation in hospital care. The Thornton family filed one such suit against the hospital where their daughter was born, and died, for "the emotional distress" they suffered by being given pictures of Hannah taken shortly after her death. She had lived for two weeks before they made the decision to remove life support. The parents had taken many photos during those weeks that she lived but objected to the posed photographs taken by hospital staff without their consent after her death. The father described the photos as "haunting to look at. She's posed as a doll or puppet. It's spooky." Another man in the Chicago area sued a hospital after seeing pictures of his dead grandchild who had been born to his teenaged daughter (Marosi, 1999). Mary Ellen Mannix (2003), a UNITE contributor, expressed similar sentiments in a piece entitled "Bringing Home a Box." Her son had been born with a congenital heart defect and lived eleven days. Mary tells of opening the green memory box, which the hospital had given her, on the way home after her son's death, and being horrified to find a photograph of her son after death that had been taken against her will. She recalls "I could not have made [my wishes] more clear to the nurse who asked" whether we would like a postmortem photo. "No. no, no. I do NOT want a picture of him like that." Apparently, after heart surgery he had become very swollen and discolored. He "had gone from a beautiful little Caucasian boy with his dad's features and my round face to a grotesquely distorted purple, blue, and black extra-large baby. I never wanted to see a photo of him like that. That was an image I already knew I would have difficulty forgetting. . . . He still looked like he was in pain. He still looked like he had been butchered for absolutely no reason."[13]

Information Technologies and Social Support

The advent of web-based support has changed the way that pregnancy loss support is now offered. The first pregnancy loss support groups were established in the United States in the mid-1970s, and during the 1980s such groups spread quickly throughout the country. By 1993, there were more than nine hundred pregnancy loss support groups, and similar groups had been established in Canada, Australia, Israel, Italy, England, West Germany, South Africa, and the Virgin Islands (SHARE, 1993). By 2000, the number of pregnancy loss support groups had dropped to just over seven hundred; of those, approximately thirty to forty did not offer peer support. The drop in numbers is due in part to the fact that people are getting better support in hospitals at the time of their loss, and this hospital-based care often includes a one-on-one follow-up session; indeed, one large organization that had offered peer support changed its policy and now offers only one-on-one support (Cathi Lammert, personal communication, July 2000).

Another important factor leading to a diminution of face-to-face, peer-based pregnancy loss support is the advent of web-based support. SHARE and UNITE each now has a web presence. In addition, there are numerous other electronic sites devoted to pregnancy loss support. Some of these include Hygeia, a site run by Dr. Michael Berman, a Yale ob-gyn who seeks to create "a global community using 'new technologies' to assuage the hurt, grieving, and sorrow experienced by families who have endured the loss of a pregnancy, newborn, or infant child" (http://www.hygeia.org, accessed January 14, 2002); "Heartbreaking Choice," a "nonjudgmental place to find support . . . for people who have chosen to terminate pregnancies due to severe fetal health problems"; Christian Pregnancy Loss Support, which offers "spiritual comfort, solace, prayer and emotional support" to "women who have suffered from miscarriage, stillbirth or neonatal death or infertility"; and listservs for "Miscarriage after Infertility," "Multiple Miscarriages," and "Pregnancy after Miscarriage."

Scholars of "computer-mediated social support" have identified a number of features or "affordances" that may make this form of peer support particularly attractive. Walther and Boyd's (2003) online survey of 340 users from 57 usenet newsgroups designed to offer support for a wide range of issues, including one devoted to pregnancy loss (soc.support.pregnancy.loss), indicates that users appreciate the greater ease of access. Computer-mediated support is available every day, around the clock, and allows access to people who are geographically distant. Computer-mediated support is also perceived as providing access to a greater range of expertise.[14] Walther and Boyd (2003) also found computer-mediated communication allowed participants to "optimize their expressiveness," noting that "participants may craft their messages and their self-presentations with greater care than may be possible in spontaneous face-to-face settings. They may stop and think, edit, rewrite, even abort and re-start a message." Because people do not need to monitor their "gestures, facial expressions, voice or physical appearance," they may pay more attention to their desired message and may have heightened "self-awareness" (p. 170). In addition, computer-mediated support offers a higher level of anonymity and social distance, which is considered beneficial because participants can be more candid and "manage stigma" with greater ease. These features also reduce the obligation to reciprocate. Unlike face-to-face support groups, where members may feel "social pressure to participate actively and disclose their thoughts and feelings" (Galagher et al., quoted in Walther & Boyd, 2003, p. 160), with computer-mediated support, people may participate as "lurkers," reading other's contributions without revealing their own presence.

Conclusions

So, what have been the effects of new medical and information technologies on the experience of pregnancy loss? As scholars of technology have come to expect, the introduction of new technologies almost inevitably results in complex changes that include unintended, often ironic, consequences. Pregnancy loss proves to be a particularly good case in point as it provides numerous examples of what Tenner (1996) describes as "revenge effects" (p. 7), effects which result when an intervention meant to reduce a negative effect actually produces another of the very type of problem it was meant to solve.[15]

Some new reproductive technologies have reduced the number of perinatal losses. Others (e.g. hormonal treatments, IVF, GIFT, ZIFT), designed to increase women's reproductive capacities, have led instead to higher rates of loss. When these losses come in the form of an ectopic pregnancy, they pose a potentially fatal risk to women and typically will permanently damage their reproductive capacities.

Prenatal diagnosis has, in many cases, simply moved up the moment of death while placing the burden of choice on would-be parents. Although (as I know firsthand) neonatal intensive care has reduced the number of late losses significantly, for some it just delays a loss while inflicting additional suffering on the child and parents, and placing the heartbreaking, end-of-life decisions on them.

Narratives of linear progress shape popular scientific accounts of prenatal development in a way that ignores and/or hides the frequency with which embryos/fetuses (and chicks) fail to develop. This narrative structure is homologous with cultural assumptions about the nature of technoscientific progress and is reinforced by science reporting of "revolutions" in reproductive medicine. The respect afforded expert knowledge—in particular, the special, culturally-specific authority that doctors hold in the United States (Starr, 1984)—coupled with the faith that many Americans have in the power of "advanced technology," contributes to unrealistic expectations about pregnancy outcomes, which in turn aggravates the experience of loss. Individuals who suffer pregnancy loss (like those who suffer job loss and downward mobility, premature birth, or other misfortunes that challenge narratives of linear progress) tend to be ill-prepared for these eventualities, as are members of their social networks (Layne, 2000). Narratives of linear progress add unnecessary angst to those involved and hamper the enactment of a host of social and technical aids for these problems.

The experience of loss is also aggravated by reproductive technologies (home pregnancy tests, fetal visualizing technologies, and prenatal tests that identify the sex) that encourage prenatal bonding. Earlier and more intensive medical management of pregnancy encourages earlier and more intensive social construction of fetal personhood in wished-for pregnancies. Home pregnancy tests make a particularly interesting case for scholars of gender and technology. They are low-tech, affordable, and easy to obtain and use—in other words, they are the very type of technology advocated by the women's health movement (and STS scholars calling for more democratic design and use of technoscience). At first glance, it appears that a home pregnancy test puts power/knowledge in the hands of women. But although these tests boast an accuracy level of 99 percent, accounts published in the support newsletters suggest that they are not authoritative (Layne in prep.). Women seek, and doctors insist on, confirmation by another test done at the doctor's office (e.g., Jennifer Fisher [2002] tells how, after finding out she was pregnant by using a test at home, she "went to the doctor for confirmation. Sure enough, I was pregnant.").[16]

Thus, "diagnosing" a pregnancy has become a two-part, technologically dependent process. It is not just that missed menses and bodily changes no longer are considered a reliable method of learning that one is pregnant. Not one, but two scientific tests are undertaken. Home diagnostic kits do not replace doctors' tests, they are just an additional, prior step, and they represent yet another instance of increasing pregnancy-related consumption (Taylor, 1998, 2000a, 2000b; Taylor, Layne, & Wozniak, 2004). Clearly, these tests are profitable for the pharmaceutical companies that produce them.

The determination of a pregnancy at such an early stage is a mixed blessing. The women's health movement is based on the premise that the more women know about their bodies, the more empowered they will be. But not all knowledge is empowering. Early diagnosis of pregnancy is certainly a good thing in cases of unwanted pregnancies. It may also be of value to women who will make changes in their daily habits upon learning they are pregnant. But it also means that women who would have in the past been spared the experience of a pregnancy loss now must deal with a loss, and do so in a culture that denies and belittles this experience.

Early detection also adds to and extends stress for women undergoing a subsequent pregnancy after a loss. Furthermore, since many women try to protect themselves from another heartbreak during a subsequent loss by trying not to become attached to the pregnancy, and given the hype about the importance of maternal–fetal bonding, these women are also subject to negative judgments and self-judgments—yet another source of strain. For example, as Kelly Gonzalez describes, "I was flooded with guilt. I did not want to punish this baby by not being as excited as I was with Alycia. But I just couldn't allow myself to become attached" (Gonzalez, 1988). Women in Cote-Arsenault, Bidlack and Humm's (2001, p. 132) study expressed the following worries: "whether I'm going to 'wreck' this baby by my fears and concern, [if my] sadness [would] hurt him," and "whether or not the baby could sense all my anxiety." One woman who had had a full-term stillbirth was made to feel guilty by her husband for her inability to read to their unborn baby during a subsequent pregnancy as she had done during the pregnancy that ended in loss. He told her, "Well, it's really not fair, you know, you have to try. I know how hard it is for you, but you have to give this baby support, too" (Cote-Arsenault & Marshall, 2000, p. 481).

One likely explanation for this comes from the patriarchal ideology of pregnancy, which is fetal- rather than woman-focused, described so cogently in the work of Barbara Katz Rothman (1993). As I point out in *Motherhood Lost,* earlier bonding is considered beneficial because it is likely to have beneficial health effects to a fetus/newborn by encouraging good health habits during the pregnancies. But negative consequences to women, although recognized in the medical literature (Cote-Arsenault & Morrison-Beedy 2001; Cote-Arsenault, Bidlack, & Humm, 2001), are easily dismissed.

As Barbara Katz Rothman observes, this fetus-centered focus on pregnancy reflects the patriarchal emphasis on production and quality assurance. Rothman (1993) describes how hard it was for people who read (or heard) about her study of amniocentesis (which "looked at the costs of these technologies to women") to understand this focus. "So what (they ask) are the long-term consequences for the child of an early diagnosis of fetal sex? So what (they ask) are the long-term consequences of the tentative pregnancy on 'bonding' and infant development? The assumption, the heart of their argument, is that if something has no consequences for the child, it has no consequences" (p. viii).

Sonograms have also come to play a central role in the diagnosis of pregnancy loss. As far as I know, the costs and benefits of this innovation to women have not been considered either in the medical or social scientific realms. Probably a large percentage of the nearly one million pregnancy losses that occur in the United States each year are diagnosed during "routine" sonograms. Yet this fact is almost entirely invisible. This is the source of irony that many contributors to the newsletters note when they report that they learned their baby had died during a "routine" sonogram exam. In this, both home pregnancy tests and the routinized use of obstetrical ultrasound and of sex identification as a standard part of prenatal diagnosis present us with a prime example of what Winner (1986) describes as "technological somnambulism" and yet another case of the tyranny of information. If there is information to be had, one is considered irresponsible not to seek it.

Yet, as Rothman has observed, the information age involves "something beyond knowledge for its own sake . . .: it is the idea that action is based on information, and the fullest possible information is needed to determine action responsibly. Informed consent is more than a legal requirement: it is a developing social norm. If there is information to be had, and decisions to be made, the value lies in actively seeking the information and consciously making the decision. To do otherwise is to "let things happen to you," not to "take control of your life" (1993, p. 83). Yet pregnancy loss is almost always about things happening to you over which you (and your medical caregivers) have no control! In cases where a demise has already occurred, there is really nothing to do. What, if any, are the advantages of knowing about an imminent abortion or stillbirth?

After a demise has been discovered, options about how, when, and where to end the pregnancy may give both patient and caregiver a measure of comfort in being able to assert some choice or control over at least one aspect of a pregnancy that has swung so devastatingly and irrevocably out of control. Even so, there are costs associated with this. Women may, for instance, do as I once did: use hormone therapy to prolong a pregnancy after demise had been determined via sonogram so that I could attend a professional conference before having the D&C. While this was certainly an effort on my part to reassert control of my life and minimize my losses, this control is not without costs—the psychic costs of dissimulation (pretending at the conference that I was fine) and self-alienation and disgust (feeling like I was carrying a rotting corpse inside). Ann Bachman (1999) tells of a similar experience of dissimulation she engaged in as the result of learning via ultrasound that one of her twins had anencephaly—"a lethal birth defect where the top of the brain and skull does not completely form." She and her husband learned this in April, the day before their daughter's First Communion and a few days before her father-in-law's eightieth birthday party. "I didn't want to spoil these very special days. I had to pretend all was well." As her pregnancy progressed, she became "hugely pregnant, obviously having twins. . . . I didn't share everything with all who asked if it was twins. So, for a month I pretended I was having healthy twins. It was a crazy world to live in. It broke my heart." Her doctor asked her if she wanted to be induced, "but I couldn't pick the day that David's life would end."

In some conditions, such as twin-to-twin transfusion syndrome, or when one fetus is missing a heart but sustained by the circulatory system of its twin (Blizzard, in press), both of which may be diagnosed by another fetal imaging technology (embryofetoscopy), families are presented with excruciating choices about terminating one fetus in order to increase the chances for survival of the other (Blizzard, in press). Here, the costs to parents of making such a choice are weighed against the benefit of improving the chances for at least one live birth. However, when sonograms (and other forms of prenatal diagnosis) reveal conditions that are likely or certainly going to result in a demise (as in Ann Bachman's case just described), women are presented with heartbreaking "choices" about how and when to end the life of a wished-for child. In these cases, there are no benefits to a potential child, and the costs of this to the woman and her partner seem to go unrecognized.

Another problem is that these consumer/medical choices about what to do after a demise has occurred are *not* presented to women until they are in crisis. This contrasts with labor and delivery, where women are educated in advance about what to expect and what their medical options will be, and they are encouraged to create a "birth plan" in order to maximize their self-determination during birth. Women are not prepared during routine prenatal visits for the 15–20 percent likelihood that their pregnancy will end in loss. Furthermore, for late losses, women are often induced and go through a full labor and delivery without having had the opportunity to attend childbirth classes.

Imaging technologies (including sonograms, embryofetoscopy, and conventional photography) as well as traces of a heartbeat on the printout of an electronic fetal monitor are now often used by the medical staff and bereaved parents as therapeutic technologies designed to confirm and affirm that a loss took place—a loss that is worthy of memory. It is unclear how much research was conducted before these innovations were routinized. A controversial study published in *Lancet* (Hughes P. et al 2002) suggests that similar caregiving practices like encouraging parents to see and hold their infant after a stillbirth, introduced in the last twenty years, may increase rates of depression. While this study is based on a very small sample and is clearly not adequate to prove that this practice is damaging, it does raise questions regarding the process through which new therapeutic technologies are adopted.

The advent of computer-based support has enhanced pregnancy loss support in a number of ways by increasing access in terms of time and geographic location. It makes it easier for people to locate and communicate with others who have had losses of a similar nature. It may be that written (and rewritten) expression may have some advantages over oral, spontaneous expression in terms of "heightened self-awareness." The greater degree of anonymity and social distance that computer-mediated communication offers is presented as a benefit of computer-mediated support because it allows participants to more easily "manage stigma." But since one of the main thrusts of pregnancy loss support movement is that pregnancy loss is *not* something one needs to be ashamed of, is *not* something one should feel must be hidden, it is unclear that the anonymity this technology offers is really beneficial. Ironically, the greater the need for anonymity, the more important it is for the problem to be brought out in the open. As Kirmayer has observed, "there is a crucial distinction between ... a public space of solidarity and a private space of shame" (1996, p. 189). Furthermore, in cases where pregnancy losses occur as part of an occupational cluster, or in toxically assaulted communities, keeping such losses hidden clearly stymies opportunities for political mobilization and community building.

Another problem with support now offered in both peer support groups and web-based support is that support is now offered after a loss occurs. If people only learn about pregnancy loss after they have one, and if the only people who know about such losses are people who have had them or been directly affected by them, then women and people who undertake a pregnancy will continue to be uninformed and unprepared for this not-uncommon eventuality. These types of support keep the topic hidden in protected real or virtual spaces.

Despite twenty years of social scientific research on the impact of new reproductive technologies, and despite the women's health movement's emphasis on empowering women through increasing their control over their bodies, a host of new technologies has significantly changed the experience of pregnancy loss in unintended *and* unexamined ways. Even in the area of reproductive medicine, where one would expect to find very low degrees of technological somnambulism given feminists' keen attention to these matters, in the case of pregnancy loss (something that effects almost one million American women each year), we seem to be in a deep, deep sleep. Take this, then, as a wake up call—"Let us sleep no more."

Notes

1. The rate of loss also increases at a similar rate with paternal age (Cunningham, 2001, p. 856).

2. The demand for regular obstetrical services had declined between 1965 and 1975 with declining fertility rates, and many doctors began to specialize in reproductive endocrinology/infertility once board certification became available in 1974 (Sandelowski, 1993, p. 9). In the 1980s and 1990s, large numbers of women underwent fertility treatments. More than one million new patients were reported to have sought treatment for infertility in the United States in 1990 (Sandelowski, 1993, p. 7). For these women and their partners, the increased level of financial, emotional, and time commitment involved in their quest for a biologically related child led to a huge investment in the pregnancy (p. 126–127).

3. Another example is found with the Spanish painter, Goya. He married Josefa Bayeu in 1773 and she is reported to have had "as many as twenty pregnancies," which resulted in the birth of seven children, only one of whom survived (Updike 2003).

4. See Layne, 2003; Conklin & Morgan, 1996; Morgan, 1996a; and Morgan, 1996b, for more on the social construction of fetal personhood.

5. They went on to discuss the uses to which ultrasound might be put in the struggle surrounding fetal rights, such as requiring that women considering abortion or refusing fetal therapy be forced by court order to view an ultrasound (1983, p. 393). Such suggestions have received the outraged condemnation they deserve (Petchesky, 1987, p. 277), but the practice has been adopted by anti-abortion pregnancy crisis centers (Leland 2006).

6. See Rapp (1997) for a discussion of its development in relation to prenatal testing.

7. Petchesky cites a study that estimated that ultrasounds were used on at least one-third of all pregnant women in the United States by 1987 (Petchesky, 1987, p. 273); Rapp cites a study published in 1993 that gives an estimate of 50 percent of pregnant women in the United States receiving ultrasounds, but she suggests that in urban areas the rate is probably closer to 90 percent (Rapp, 1997, pp. 31,48); and Taylor (199, p. 18) cites a national survey that reports "between 1980 and 1987, the percentage of all pregnancies that were scanned by ultrasound in the United States increased from 35.5 percent to 78.8 percent." Prenatal scanning has also become common in Canada (Mitchell, 1994, p. 148), Australia, and much of Europe (Mitchell & Georges, 1998; Georges, 1996; Satenena, 1996 and 2000).

8. Of course, sex identification does not always lead to more intense attribution of personhood. In countries like India (Kristof 1993) and China (Rohde 2003), ultrasounds are being used so that female fetuses may be aborted. See Mitchell (1994) for a discussion of the controversy in Canada over the use of ultrasound for fetal sex identification.

9. Research on subsequent pregnancies suggests that other women have similar experiences. According to Cote-Arsenault, Bidlack, and Humm (2001b), women report that "worries never end." Some told of being "'scared to bring the baby home,' . . . [they] realized that additional calamities could occur; one woman said that she worried about SIDS. . . . The end of pregnancy is not truly the end of concerns because their confidence in the world has been shattered" (pp. 132–133).

10. The Centering Corporation publishes a manual on how to take photographs of stillborns and infants who die (Johnson et al., 1985).

11. Ninety-five percent of the respondents to the SHARE survey "felt that it is important to have pictures of the baby" (Laux, 1988a). If the parents do not want the photos, the hospital often keeps the photos for some time in case the parents change their mind.

12. I have no evidence of bereaved parents using sonogram images for this purpose after a death, but Taylor (2000a), and Mitchell (1994) report North American women engaging in the practice during real-time sonogram exams.

13. The benefits of seeing and holding stillborn babies is also being challenged in some quarters. A British study published in *Lancet* (Hughes P. et al. 2002), based on research on sixty-five women from the United Kingdom, suggests that women may not benefit from seeing and holding stillborn babies (http://www.nelh.nhs.uk/hth/stillbirth.asp). See the September/October 2002 SHARE newsletter (11 [5], p. 3) for a discussion of this study and the pregnancy loss support movement's response.

14. In the case of pregnancy loss, such sites allow bereaved parents to locate not just other bereaved parents, but more specifically parents whose child died under similar circumstances or due to similar causes. This function is also accomplished through the newsletters. The SHARE newsletter includes a "parent connection" section where parents would ask to be contacted by others who had had similar experiences, e.g. "anyone who has experienced a loss due to incompetent cervix," or "anyone whose baby died of a rare heart condition, subendocardial fibrosis," or "anyone who experienced a stillbirth, and then went on to have another baby" (SHARING, 1999, 8[1], p. 7). The SHARE office serves as an intermediary in these cases—i.e., someone wanting to respond to such a solicitation contacts the SHARE office for the person's contact information.

15. As Tenner notes, "Technology alone usually doesn't produce a revenge effect. Only when we anchor it in laws, regulations, customs, and habits does an irony reach its full potential. . . . Revenge effects happen because new structures, devices, and organisms react with real people in real situations in ways we could not foresee" (1996, p. 9).

16. Furthermore, each company also offers what the computer industry would call "support service." For example, if (after following the manufacturer's "friendly instructions" in either English or Spanish) one still has questions, e.p.t.'s maker Pfizer, for example, has registered nurses available at a toll free number during business hours and provides recorded assistance twenty-four hours a day.

References

Becker, Gay. (2000). *The elusive embryo: How women and men approach New reproductive technologies.* Berkeley: University of California Press.

Blizzard, Deborah. (in press). *Looking within: A sociocultural examination of fetoscopy.* Cambridge: MIT Press.

Budetti, Peter, Barrand, Nancy, McManus, Peggy, & Heinen, Lu Ann. (1981). The costs and effectiveness of neonatal intensive care (Background paper #2). In *Case studies of medical technologies, case study #10*. Washington, D.C.: Office of Technology Assessment.

Burgan, Renee M. (1988). To my son, William Randolph Burgan. *UNITE Notes* 7(2), 4.

Brunner, Donna L. (1992). Me? Guilty? *UNITE Notes* 11(2), 3.

Caiola, Julie. (2000). To Angel. *UNITE Notes* 18(3), 4.

Cartwright, Lisa. (1993, November). Gender artifacts in medical imaging: Ultrasound, sex identification, and interpretive ambiguity in fetal medicine. (Paper delivered at the American Anthropological Association's annual meeting in Washington, D.C.)

Casimer, Lisa A. (1987). Sarah's story—with love. *SHARE Newsletter* 10(3), 11–13.

Coggins, Chrissy. (2000). God doesn't make mistakes. *UNITE Notes* 18(4), 7.

Cohen, Elizabeth. (1997, May 4). The ghost baby. *New York Times Magazine*, 84.

Cole, Catherine. (1993). Sex and death on display: Women, reproduction, and fetuses at Chicago's Museum of Science and Industry. *Drama Review 37*, 1, 43–60.

Cooley, Anna. (2002). This is to my baby. *SHARING* 11(2), 1–2.

Cote-Arsenault, Denise. (2000). One foot in—one foot out: Weathering the storm of pregnancy after perinatal loss. *Research in Nursing & Health 23*, 473–485.

Cote-Arsenault, Denise, Bidlack, Deborah, & Humm, Ashley. (2001). Women's emotions and concerns during pregnancy following perinatal loss. *The American Journal of Maternal Child Nursing 26*(3), 128–134.

Cote-Arsenault, Denise, & Morrison-Beedy, Dianne. (2001). Women's voices reflecting changed expectations for pregnancy after perinatal loss. *Journal of Nursing Scholarship*, (3), 239–244.

Cote-Arsenault, Denise, & Dombeck, Mary-T. B. (2001). Maternal assignment of fetal personhood to a previous pregnancy loss: Relationship to anxiety in the current pregnancy. *Health Care for Women International 22*, 649–665.

Couto, Richard A. (1986). Failing health and new prescriptions: Community-based approaches to environmental risks. In Carole E. Hill (Ed.), *Current health policy issues and alternatives* (pp. 53–70). Athens: University of Georgia Press.

Cunningham, F. Gary, Gant, Norman F., Leveno, Kenneth J., Gilstrap, Larry C. III, Hauth, John C., & Wenstrom, Katharine D. (2001). *Williams obstetrics* (21st Ed.). New York: McGraw-Hill.

Cuce, Barbara. (1998). Somewhere you haven't been. *UNITE Notes* 17(2), 6.

Dowling, Claudia Glenn. (1993, December). Miracles of birth: The blessings of a medical revolution; Healthy babies who 10 years ago would never have been born. *Life*, 75–77.

Down, Martyn Langdon. (1986). *Some babies die: An exploration of stillbirth and neonatal death* [Film]. Australia: University of California Extension, Center for Media and Independent Learning.

Doyle, Emma. (2002). My Angel. *UNITE Notes 21*(2), 4.

Dumit, Joseph (with Sensiper, Sylvia). (1998). Living with the 'truths' of DES: Toward an anthropology of facts. In Robbie Davis-Floyd and Joseph Dumit (Eds.), *Cyborg babies: From techno-sex to techno-tots* (pp. 212–239). New York: Routledge.

Evans-Smith, Heather Gail. (2002). Fear. *UNITE Notes 21*(2), 3.

Fadako, Rita. (1997). Pictures of an angel. *UNITE Notes 16*(1), 1.

Fillion, Kate. (1994). Fertility rights, fertility wrongs. In Gwynne Basen, Margrit Eichler, & Abby Lippman (Eds.), *Misconceptions: The social construction of choice and the new reproductive technologies* (vol. 2) (pp. 33–55). Ontario: Voyageur Publishing.

Fisher, Jennifer K. (2002). This is the story of our daughter, Catalina Pearl Robertson. *SHARING 11*(5), 6.

Fletcher, John, & Evans, Mark. (1983). Maternal bonding in early fetal ultrasound examinations. *New England Journal of Medicine 308*(7), 392–393.

Friedrich, Jim. (1984). Reflections without mirrors. *SHARE Newsletter 7*(3), 2.

Gainer, Julie. (2000). My beautiful Aidy. *SHARING 9*(2), 7.

Halpern, Sue. (1989, September). *Ms,* 56–64.

Heil, Janis. (1982). Giving birth to death. *UNITE Notes 2*(1), 2.

Hicks, Dana, & Hicks, Travis. (2001). Our story: In memory of Lance and Chase Hicks. *SHARING 10*(6), 6.

———. 1988b. A miscarriage hurts, too. *UNITE Notes 8*(1), 3.

———. 1988c. Two years. *UNITE Notes 7*(3), 4.

Horning, Anita. (1997). Eight priceless days. *SHARE Newsletter 6*(2), 15.

Hughes, P. et al. (2002). Assessment of guidelines for good practice in pschosocial care of mothers after stillbirth: A cohort study" *Lancet* Jul 13; 360(9327):114–18.

Iacono, Linda. (1982). Faith. *UNITE Notes 2*(1), 4–6.

Ingle, Kristen. (1981/82a). Pink blankets. *UNITE Notes 1*(2), 2.

———. (1981/82b). For Elizabeth at Christmas. *UNITE Notes 1*(2), 1.

———. (1986/87). Subsequent pregnancy. *UNITE Notes 6*(2), 4–5.

Johnson, Joy, and Dr. S. Marvin Johnson, Chaplain James H. Cunningham, Irwin J. Weinfeld, M.D. (1985). *A most important picture: A very tender manual for taking pictures of stillborn babies and infants who die.* Omaha, NE: Centering Corportation.

Jones, Carl. (1987). Fiery tale. *SHARE Newsletter 10*(2), 3.

Kiper, Suzanne. (2002). In the pain of stillbirth, there is still hope. *SHARING 11*(2), 6.

Kirmayer, Laurence J. (1996). Landscapes of memory: Trauma, narrative, and dissociation. In *Tense past: Cultural essays in trauma and memory,* Antze and Lambek eds., pp. 173–198. New York: Routledge.

Kristof, Nicholas D (1993). The chosen sex—A special report; Chinese turn to ultrasound, scorning baby girls for boys. *New York Times,* July 21.

Lasker, Judith N., & Borg, Susan. (1987). *In search of parenthood: Coping with infertility and high-tech conception.* Boston: Beacon Press.

Launslager, Donna. (1994). From 1–5: Multiple births and the family. In Gwynne Basen, Margrit Eichler, Abby Lippman, (Eds.), *Misconceptions: The social construction of choice and the new reproductive technologies* (vol. 2) (pp. 117–126). Ontario: Voyageur Publishing.

Laux, Jana. (1988a). SHARE baby pictures questionnaire—part I. *SHARE Newsletter 11*(3), 10–13.

———. (1988b). Baby questionnaire—part II. *SHARE Newsletter 11*(4), 10–12.

Layne, Linda L. (1992). Of fetuses and angels: Fragmentation and integration in narratives of pregnancy loss. In David Hess & Linda L. Layne (Eds.), *Knowledge and society: The anthropology of science and technology* (vol. 9) (pp. 29–58). Greenwich, Conn.: JAI Press.

———. (2001, Spring/Summer). The search for community: Tales of pregnancy loss in three toxically assaulted communities in the U.S. [special issue]. *Women's Studies Quarterly*, 25–50.

———. (2003). *Motherhood lost: A feminist account of pregnancy loss in America*. New York: Routledge.

———. (2006). Optimizing care during subsequent pregnancy: A conversation with Denise Cote-Arsenault, Ph.D., R.N. and Pam Scarce, R.N. An episode of the educational television series *Motherhood Lost: Conversations.* Co-produced with Heather Bailey, Fairfax, Va.: George Mason University Television.

———. (in press). 'A women's health model for pregnancy loss': A call for a new standard of care. *Feminist Studies.*

———. (in prep.). The home pregnancy test: A feminist technology? In *Feminist Technology?* Eds. Linda Layne, Kate Boyer and Sharra Vostral. Book manuscript.

Leland, John (2006). Some abortion foes forgo politics for quiet talk. *New York Times.*

Lillyman, Anne-Marie. (2001). A test of faith. *SHARING 10*(4), 1.

Mannix, Mary Ellen. (2003). Bringing home a box. *UNITE Notes 23*(3), 4.

Marosi, Richard. (1999, July 5). Mourning baby with a snapshot in death. *Los Angeles Times*, A1, A18.

Miller, Branda. (1994). *Witness to the future* [Film]. Troy, N.Y.: Iear Studio, RPI.

Mitchell, Lisa M. (1994). The routinization of the other: Ultrasound, women and the fetus. In Gwynne Basen, Margrit Eichler, & Abby Lippman (Eds.), *Misconceptions: The social construction of choice and the new reproductive technologies* (vol. 2) (pp. 146–160). Ontario: Voyageur Publishing.

Mitchell, Lisa M., & Georges, Eugenia. (1998). Baby's first picture: The cyborg fetus of ultrasound imaging. In Robbie Davis-Floyd & Joseph Dumit, (Eds.), *Cyborg babies: From techno-sex to techno-tots* (pp. 105–124). New York: Routledge.

Nelkin, Dorothy. (1995). *Selling science: How the press covers science and technology* (Rev. ed). New York: Freeman and Co.

Niles, Tara. (1996). The missing piece in our three-piece puzzle. *SHARE Newsletter 5*(1), 8.

Nixon, Rob. (2003, April 13). [Review of the book *Things my mother never told me*]. *New York Times Book Review*, 6.

Office of Technology Assesssment (1987). *Neonatal intensive care for low Birthweight infants: Costs and effectiveness*. Health Technology Report 38. Washington D.C.

Pizer, Hank, & Palinski, Christine O'Brien. (1980). *Coping with a miscarriage: Why it happens and how to deal with its impact on your family*. New York: Signet.

Rapp, Rayna. (1997). Real-time fetus: The role of the sonogram in the age of monitored reproduction. In Downey and Dumit (Eds.), *Cyborgs & citadels: Anthropological interventions in emerging sciences and technologies* (pp. 31–48). Santa Fe, N.M.: School of American Research.

———. (1999). *Testing women, testing the fetus: The social impact of amniocentesis in America*. New York: Routledge.

———. (2001). Gender, body, biomedicine: How some feminist concerns dragged reproduction to the center of social theory. *Medical Anthropology Quarterly* 15(4), 466–477.

Rohde, David (2003). India steps up effort to halt abortions of female fetuses. *New York Times*, October 26.

Rothman, Barbara Katz. (1993). *The tentative pregnancy: How amniocentesis changes the experience of motherhood*. New York: W. W. Norton.

Sandelowski, Margarete. (1993). *With child in mind: Studies of the personal encounter with infertility*. Philadelphia: University of Pennsylvania Press.

SHARE. (1993). *International perinatal support groups*. Belleville, Ill.: St. Elizabeth's Hospital.

Simons, Cari. (1995). The anniversary. *SHARE Newsletter* 4(5), 7.

Tangir, Jacob, M.D. (2001). Infection and pregnancy loss. In Michael Berman (Ed.), *Parenthood lost: Healing the pain after miscarriage, stillbirth, and infant death*. (pp. 168–179). Westport, Conn.: Bergin & Garvey.

Taylor, Janelle S. (1998). Image of contradiction: Obstetrical ultrasound in American culture. In Sarah Franklin and Helena Ragone (Eds.), *Reproducing reproduction: Kinship, power, and technological innovation* (pp. 15–45). Philadelphia: University of Pennsylvania Press.

———. (2000a). An all-consuming experience: Obstetrical ultrasound and the commodification of pregnancy. In Paul Brodwin (Ed), *Biotechnology and culture: Bodies, anxieties, ethics* (pp. 147–170). Bloomington: Indiana University Press.

———. (2000b). Of sonograms and baby prams: Prenatal diagnosis, pregnancy, and consumption. *Feminist Studies* 26(2), 391–418.

Tenner, Edward. (1996). *Why things bite back: Technology and the revenge of unintended consequences*. New York: Alfred. A. Knopf.

Treichler, Paula A., Cartwright, Lisa, & Penley, Constance. (1998). Introduction: Paradoxes of visibility. In Paula A. Treichler, Lisa Cartwright, Constance Penley (Eds.), *The visible woman: Imaging technologies, gender and science* (pp. 1–17). New York: New York University Press.

Updike, John. (2003). An obstinate survivor: Robert Hughes takes on the life of Goya. *New Yorker* November 3. 88–91.

Ventura, Stephanie J., Mosher, William D., Curtin, Sally C., & Abma, Joyce C. (2001). Trends in pregnancy rates for the United States, 1976–97: An update. *National Vital Statistics Reports 49*(4).

Ventura, Stephanie J., Taffel, Selma M., Mosher, William, Wilson, Jacqueline B., & Hershaw, Stanley. (1995, May 25). Trends in pregnancies and pregnancy rates: Estimates for the United States, 1980–92. *Monthly Vital Statistics Report 43*(11). Centers for Disease Control and Prevention/National Center for Health Statistics.

Walther, Joseph B, and Boyd, Shawn. (2003). Attraction to computer-mediated social support. In C. A. Lin & D. Atkins (Eds.), *Communication technology and society: Audience adoption and uses of the new media* (pp. 153–187). New York: Hampton Press.

Weber, David. (2001). Grace and faith. *SHARING 10*(6), 1–2.

White, Tami. (2000). Choices we made. *SHARING 9*(2), 1–2.

Winner, Langdon. (1986). *The whale and the reactor: A search for limits in an age of high technology.* Chicago: University of Chicago Press.

8

Feminist Narratives of Science and Technology: Artificial Life and True Love in *Eve of Destruction* and *Making Mr. Right*

CAROL COLATRELLA

Although many social scientists studying the digital divide and the underrepresentation of women and minorities in scientific and technical fields consider how gender, race, class, and other cultural differences affect access to and engagement with science and technology (Fox, 1999; Etzkowitz, Kemelgor, & Uzzi, 2000; Rosser, 1997), fewer humanists, who are generally more interested in aesthetics rather than ethics, discuss such issues (Balsamo, 1996; Baym, 2001). Since feminine use of technology is more often depicted in popular fiction, especially science fiction, rather than high-culture narratives, it is not surprising that woman's engagement with technoscience has attracted greater attention from cultural studies critics. Narrative representations and cultural values of gender reinforce each other; therefore, it is critically important that social and textual analyses converge and create "crossover" between disciplines concerned, respectively, with practices and representations (Oldenziel, 2001, p. 144).

Drawing on feminist social studies of science and technology and incorporating cultural studies methodologies, this chapter compares two film narratives commenting on the capacities of two female characters—one a scientist and one not—to develop experimental technologies designed to improve human life. Both films depict robots created or deployed by women and incorporate cultural stereotypes describing women's work in masculine domains of science and technology that are familiar to anthropologists and sociologists studying universities and corporations (Fox, 1998, 2001; McIlwee & Robinson, 1992). *Eve of Destruction* (Gibbins, 1991) and *Making Mr. Right* (Seidelman, 1987) employ narrative conventions associating science with self-aggrandizement, ambition, and aggression, traits identified with men, while suggesting that in some near future, feminist scientists (or perhaps even women who are not scientifically trained) might have more beneficial influence on science and on its resulting technological products.

That women ought to be involved for ethical, social democratic reasons in the construction and application of science and technology is a position argued by Sandra Harding and other social scientists (e.g., Oblepias-Ramos, 1991/1998) studying the gender disparities that keep women out of these disciplines and away from decision-making roles in scientific and technological fields: "Feminist science studies has come to have a central voice in the contemporary global project of redefining the terms of democracy. At its best, it has insisted that the democratic development of science and society requires that those who bear the consequences of science and technology policy decisions have a proportionate share in making them—in short, that women's interests and values count" (Harding, 2001, p. 302).

Suggesting thematically that women's participation changes how technologies are created, deployed, and understood, *Eve of Destruction* and *Making Mr. Right* hint at future success for the incorporation of feminist practices. In terms of characterization, emplotment, and tone, the films explore how cyborgs blend the human and the artificial and outline how women's contributions to technologies provide opportunities to reconfigure technological processes and products for social good.

Eve of Destruction and *Making Mr. Right* rely on cultural stereotypes of gender in depicting masculine and feminine engagement with technology, specifically in representing the design of the cyborg and its social relations with the humans around it. Although the tropes of destruction and creation incorporated into their titles indicate their different genres (thriller and romance), both films establish women as less able than men to control technologies, even those designed by women. *Eve*'s research scientist is punished for tinkering with robotic technology, while *Mr. Right*'s marketing consultant is rewarded for humanizing the robot; yet in both cases, the processes of developing and deploying technology are seen as ultimately residing outside feminine authority, for cyborgs are neutralized or empowered beyond the control of these female characters.

With roots in Western European fantasies of monsters, the late-twentieth-century cyborg has been theorized by cultural theorist Donna Haraway

(1998/1991/1985) and characterized in a variety of popular fictions and films, most notably Ridley Scott's *Blade Runner* (1982), William Gibson's *Neuromancer* (1984), and Marge Piercy's *He, She, and It* (1991/1993). Creating the cyberpunk genre, Gibson's and Scott's characters meld human and machine in dark worlds fraught with ambiguity and danger; both the novel and the film describe a future full of high-tech gadgets and resistant to human sympathy. Piercy's feminist science fiction novel more optimistically parallels two characters, an early modern figure of a golem created to protect a Jewish community in seventeenth-century Prague and a futuristic man-machine designed to protect a modern liberal enclave battling corporate hegemony in the twenty-first century; in the latter plot, two scientists, one male and one female, collaborate to create the robot Yod, whose capacities blend superior defensive skills and sensitivity. Yod becomes the much-loved paramour of a female artificial intelligence researcher (Shira Shipman) before he is sacrificed by his creators, just as his historical predecessor, the golem Joseph, is also destroyed. Piercy's novel concludes when Shira accepts that it was wrong to send a human-like machine to his death to protect others, and that she should therefore give up any hope of replacing Yod.

Instead of eschewing cyborg technology, Haraway takes a different position from Piercy's scientists in claiming that it is impossible to limit cybernetic experimentation, for it is already inextricably embodied in us and embedded in our notions of what it means to be human at the present time. Defining cyborg as "a cybernetic organism, a hybrid of machine and organism, a creature of social reality as well as a creature of fiction," Haraway's "A Cyborg Manifesto" (1998/1985/1991) testifies to "an ironic political myth faithful to feminism, socialism, and materialism" (p. 434). Haraway asserts, "By the late twentieth century, our time, a mythic time, we are all chimeras, theorized and fabricated hybrids of machine and organism; in short, we are cyborgs. The cyborg is our ontology; it gives us our politics. The cyborg is a condensed image of both imagination and material reality, the two joined centers structuring any possibility of historical transformation" (p. 435).

Haraway's argument—that technology and imagination structure cultural ideas of human identity—synthesizes principles with a long history in science and technology studies, particularly feminist studies of technology. Donald MacKenzie and Judy Wajcman (1999) summarize relevant developments in technology studies:

> By 1985, state-of-the-art research in the history of technology . . . was already predicated on the assumption that technology and society were bound

together inextricably and that the traffic between the two was emphatically two-way. Since 1985, this perspective has spread to the social sciences. The social shaping of technology, which in the mid-1980s still had something of the excitement of heresy, has now become almost an orthodoxy. . . . In the more sophisticated seminar rooms, reference to "technology" or "society" now prompts the immediate response that to talk of them as distinct entities is misleading. (p. xv)

Of particular significance is the growing prominence of feminist analysis in technology studies. Interdisciplinary in style and activist in intention, feminist scholarship in the 1970s and 1980s transformed the historical study of technology to consider the engagement of users as well as designers (Stanley, 1983/1998; Cockburn & Ormrod, 1993; Pursell, 2001). Ruth Oldenziel (2001) describes early twentieth-century technological development by identifying other actors in addition to designers: "Women as individuals and those organized into groups were part and parcel of this newly created landscape, helping to initiate or frame new meanings for new products in some cases and rejecting products in others" (p. 137). Incorporating the responses of users broadens the social study of any technology by widening the focus of analysis in ways that include the participation of women.

Feminist historians of technology now routinely incorporate Ruth Schwartz Cowan's analysis stemming from her 1976 question "Was the female experience of technological change significantly different from the male experience?" (Pursell, 2001, p. 115) and her outlining "four areas where one might look at how: with respect to women's bodies and their functions, in work outside the home, in domestic labor, and in the ideological realm" (p. 115). Parsing "the ideological realm" of technology, Judy Wajcman (1991) cogently explains "the ideology of masculinity . . . has this intimate bond with technology" in that sexuality, control, combat, and heroism are recurring themes in technological rhetoric and that there is a "continuing male monopoly of weapons and mechanical tools" (pp. 141, 150). Wajcman notes the interdependence of gender concepts: "Masculinity and femininity are produced in relation to each other and what is masculine, according to the ideology of sexual difference, must be the negation of the feminine" (p. 158).

Feminist cultural critics similarly examine how narrative representations of growing up female reflect ideologies concerning appropriate roles and behaviors for women. Marina Warner's (1994) book on fairy tales surveys representations of women in diverse tales, situating a number of Cinderella stories, for example, within the social historical context of how different generations of women economically and legally dependent on men were forced to get along in the same household (pp. 238–239). In her popular book *Backlash*, Susan Faludi (1991) includes chapters on 1980s Hollywood television shows and films to demonstrate how executive decision makers in television networks and film companies resisted positive representations of feminism and colluded in offering media saturated with conservative depictions of women.

Although representations of women and those of scientists and technological experts remain marginalized in popular visual media, special effects animation in television and film productions and increasing consumer familiarity with video games, computer simulations, and technological hardware have increased the likelihood that television and film fictional narratives frequently make use of science and technology subjects, including aeronautics, biotechnology, computing, and medicine. Because American social mores have changed to accept the possibility that women can enter these fields, many media productions include at least a token representation of women. Recent documentary and fiction films and television shows describe female characters working as professional scientists, usually as medical caregivers and researchers playing minor roles in ensemble television dramas such as *ER, CSI, Crossing Jordan,* or *Law and Order* or even in action adventure films like *Men in Black* and *I, Robot.* Fictional female scientists often offer a "different" (often personal, emotional, intuitive, or even supernatural) perspective that proves to be a useful method to resolve scientific problems and create new technologies, but frequently these softer approaches appear insufficient or flawed rather than exemplary.

Recent representations of women scientists and technology differ substantially from earlier literary examples characterizing how women engage with science and technology, for science and technology have historically been envisioned in literature and film as male domains that include women as subjects, not authorities. A number of nineteenth-century fictional works rely on classical myths in troping science as a masculine project with dangerous and even deadly outcomes for women. That sexually stereotyped human and android characters might represent science as a dangerous and sometimes deadly intervention appears to date at least from Mary Shelley's *Frankenstein* (1818/1991), which depicts Victor Frankenstein working on the risky, often derided project of creating life. Subtitled "The New Prometheus," the novel emplots science and its technological products as failed heroic endeavors doomed by masculine ambition to overreach natural boundaries. Victor eventually regrets his technological experiment and dies trying to destroy his creature. It is his "monster" who embodies domestic virtues and becomes more sympathetic to many readers because he looks for love, companionship, family, and security rather than the scientific fame sought by Victor.

Referring to the Pygmalion myth rather than the story of Prometheus, Nathaniel Hawthorne also outlines the dangers of scientific ambitions and technological tinkering in stories such as "Rappaccini's Daughter" (1844/1987) and "The Birth-mark" (1843/1987). Hawthorne's plots explore how male scientists fuel romantic passion in designing experiments with wives or daughters as victimized human subjects. Auguste Villiers de L'Isle-Adam's mechanical fantasy *L'Eve future* (1886/1982) focuses on a modern Pygmalion character who applies scientific knowledge to engineer a Galatea, finding that even an artificial woman has needs surpassing his scientific and technological ingenuity. These accounts warn readers that egoistic male scientists inevitably produce technologies that diminish rather than enhance human life. In all too many fictional works and films of the nineteenth and twentieth centuries—even those representing women as scientific or technological experts or in those narratives including feminist principles—science and technology remain masculinized pursuits seeking to control a feminized nature (Merchant, 1990).

The rare feminist tale of science and technology offers a more positive vision of how science and its applications might improve particular users and society. Lydia Maria Child delineates how ethical motivations inspire the creation of scientific knowledge and demonstrates how technology can be applied to effect social improvement. In her short story "Hilda Silfverling: A Fantasy" (1845/1997), Child depicts a conflict between scientific knowledge and domesticity, but she optimistically resolves it by technological means when the title character is preserved by a chemist experimenting with cryogenics rather than being executed for a crime she did not commit. Hilda is saved by the new freezing technology and awakes one hundred years after her trial and punishment to find a home and family in a society friendlier to women. "Hilda Silfverling" offers a feminist and positivist vision of science and technology contrasting with *Frankenstein*'s warning of how technological innovations inevitably result in unpredictable damage. In her short romance, Child imagines how a new technology allows one woman to escape from the constraints of her original patriarchal culture by alighting in a more progressive age.

Hollywood films tend to recall Shelley's fears concerning hubristic scientific theory and technological experiment rather than Child's optimistic rendering of refrigeration as progressive social engineering. Cinematic science remains akin to the supernatural, taking advantage of the fearsome visuals of mysterious laboratories and their odd mechanical products. Differently resisting the emplotment of science as a masculine endeavor with dangerous outcomes for women and families, some recent films play cute in representing the softer, gentler side of science in imaging women as scientists who are engaging, attractive, passionate, and intelligent. Daryl Hannah's astronomer in *Roxanne,* Meg Ryan's mathematician in *IQ,* and Jodie Foster's radio astronomer in *Contact* come to mind as characters softening the image of scientists and science; however, these female scientists move on narrative trajectories away from science as they move toward fulfillment of desire. Films describing information technologies (*Desk Set, Disclosure, The Net, Sleepless in Seattle, You've Got Mail*) represent such innovations as enabling romance and/or domesticity, as if the only good technologies for women are those reinforcing traditional gender roles (Colatrella, 2001).

Other Hollywood movies represent the feminine influence on science and technology as the ethical perspective and even as a feminist project intended to thwart the corrupt evils of capitalism. For example, the popular cartoon character Inspector Gadget, who appears in the 1999 film by the same name, was created by a talented female scientist who supplies a heroic police officer with new body parts. Like the Robocop and Terminator movies, *Inspector Gadget* realistically depicts the roots of science as "a world without women" (Noble, 1992). The film visualizes how the rare woman scientist confronts the hostility of male peers who mock her because she sees the cyborg as more human than machine. The female scientist suggests that technological reasons for experimenting with cyborgs are insufficient, unethical, and cruel in their outcomes, an argument that the film sets within the gendered conflict between scientists: feminine ideas about scientific practice and the purposes of technology are represented as humane, in being empathetic to individuals and ethically designed for the public good versus the arrogant, selfish profiteering pursuit of knowledge and the exploitation of technology fostered by masculine science. Science fiction genre conventions that rely on sex stereotypes are common in narratives about artificial intelligence and alien life forms. For instance, in Piercy's *He, She, and It,* it is the female scientists Malkah and Shira Shipman who shape the cyborg to incorporate sensitivity and social empathy. Certain characteristics representing the essentially feminine, including being kind, caring, nurturing, and nonviolent, enhance the socioethical capacities of the cyborg or alien while diminishing the physical stature and cultural capital of the creature (think of the robot on *Red Dwarf* or Dr. Spock on *Star Trek*). Whether motivated by romance or morality, female scientists appear more "woman" than "scientist" in most films.

Even when films allude to feminist styles of scientific and technological inquiry, these narratives resist endorsing such radical reconfigurations and end up retreating into comfortable gendered stereotypes that put women outside the realms of science and technology. Describing how a female robot powerfully reacts against the sexism directed toward her creator and other women, *Eve of Destruction* reconfigures the Frankenstein plot as a tale of a female scientist (played by Renée Soutendijk) balancing work and family. Dr. Simmons replicates herself in developing a military weapon in human shape, a robot named Eve VIII. Eve VIII has been designed to incorporate many human physical and mental characteristics so that she can exercise careful judgment in functioning as an undercover cybercop: she is capable of commenting on the quality of fine tailoring and is presumed to be competent to evaluate how much force to apply in particular circumstances. As a modern Cinderella story of a motherless female scientist who recreates herself in machine form to overcome patriarchal oppression, *Eve of Destruction* concentrates on empowering its female protagonists by technological, rather than romantic, means. Yet as the Cinderella intertext implies, this is not a strictly feminist story. In alluding to Villiers de L'Isle-Adam's *L'Eve future* (1886/1982), Eve VIII remains the locus of male desire in that she is constructed to be a mechanically perfect helpmate, albeit one who will serve military rather than domestic needs. The story of her deployment exemplifies the oppression and eventual destruction of woman in a patriarchal environment and in terms of this theme and the episodic construction of its reinforcement resembles the popular movie *Thelma and Louise.*

The robot Eve physically and emotionally resembles her attractive scientist creator, for her creator's history has been embedded in her circuitry. Because the robot shares her creator's memories of pain and pleasure, she reacts to situations as Dr. Simmons would: becoming angry when she is sexually harassed and irritated when someone interferes with family time with her son. The robot also exhibits concern for an injured colleague and indulges in tender maternal impulses in looking for Dr. Simmons's son. But, while the doctor moderates the expression of her desire by repressing her emotions and any possible radical actions, there is no check on the robot, which has no "off" switch, as the military operative called in by the Pentagon to work with Dr. Simmons to find and neutralize Eve VIII remarks.

Recruited to destroy what Dr. Simmons has created, the doctor's ally McQuade is represented as a no-nonsense, cynical, and bluntly spoken operative who would rather act than speak and who has great suspicions about elitist scientists, especially female ones. It is ambiguous whether the filmmakers deliberately link McQuade's forthrightness to ethnicity: because the African American actor Gregory Hines plays the part, racial stereotypes are both replicated and resisted. McQuade appears a consummate professional who is expert in his field of military operations, but his hair-trigger temper and his tendency to make off-color remarks flare frequently during the time he and the doctor chase the robot. While Dr. Simmons is restrained and rational, McQuade is explosively angry about the situation the doctor's research and repressed personality have created. The film is rather strained in depicting how their temperaments might influence and complement each other without looking closely at McQuade's psychology, only briefly acknowledging his distrust of science, authority figures, and women, a combination of feelings that suggests a conflict with a mother figure rather than a reaction to racism.

Only woman's flawed psyche is dissected here, for, as Dr. Simmons quickly recognizes, Eve VIII is on a mission to rectify the doctor's past by actively confronting abuses she has endured at the hands of a drunken parent, a hostile boss, and an aggressive lover. Eve VIII strikes with deadly force against those who harmed Dr. Simmons as well as anyone attempting to hurt the robot or those around her. Relying on the movie psychology of Alfred Hitchcock's *Marnie,* the film employs flashbacks to Dr. Simmons's childhood to develop a basis for the robot's violent actions. At root in Eve VIII's consciousness is the traumatic memory of young Eve witnessing how her drunken father caused her mother's death by inadvertently pushing her into the path of car. The robot hunts down Dr. Simmons's father, who has been living under an assumed name, and seeks to destroy him, veering away from achieving this goal only because Dr. Simmons distracts her. The robot's next mission, to spend more time with the doctor's son Timmy, threatens the boy and his father.

Eve VIII's apocalyptic capabilities provide a suspenseful, although long-winded, climax to the film. As the twenty-four-hour clock is about to end the robot's campaign of violence by exploding her and anyone in her vicinity, McQuade and Dr. Simmons follow Eve VIII and her hostage Timmy into the New York City subway. McQuade fires a shot into the robot's left eye, but only manages to disable her before he is injured. Recognizing that he must rely on the doctor to complete the task of eliminating the robot, McQuade throws his gun to Dr. Simmons so that she can finish the job. Dr. Simmons, who has eschewed violence throughout most of the movie, takes up the challenge of destroying her creation to protect her son. But it takes more than a gun to stop the robot. The chase ends only after Dr. Simmons strikes Eve VIII with deadly force, sticking a knife into one of the robot's vulnerable areas, her right eye. The film's narrative logic dictates that the robot can only be eliminated after her creator releases her repressed anger. Although it might appear that the doctor confronts a no-win scenario in choosing between the robotic version of herself, a technical creation, and her son, her biological issue, the film presents the female scientist as a too-intellectual character who must become more typically feminine in order to be successful. Like Piercy's fictional scientist Shira Shipman, who gives up on love and research to protect her community, Dr. Simmons destroys her creation to save New York, and, presumably, the world. She is redeemed by her psychological breakthrough, although her research project must be sacrificed. The film's explicit message: it is better to be a nurturing mother than an innovative inventor.

In acting on her creator's repressed anger, Eve VIII triggers male fears of Julia Kristeva's (1981) phallic woman and, therefore (within the film's world-view) must be eliminated. After the robot, with a bomb installed, escapes the military brass who commissioned its deployment, the female scientist and McQuade must destroy Eve VIII to protect Eve's son and the world. Eliminating Eve VIII allows Dr. Simmons to keep up her end of the government contract, but she has transformed herself into a vigilante using any weapon to protect the citizenry, problematizing whether women should engage in masculine ways with technology. This heroic transformation of the female scientist depends on her acting out of revenge, a stereotype associated with the female psyche. *Eve of Destruction* hints at a future feminist technoscience in offering the possibility of an attractive cybercop designed by a woman to defuse situations without violence, while emplotting Eve VIII's development, deployment, and destruction as the near apocalyptic end for humanity. The robot's animation must be reversed for the world to return to a stable state; it is a weapon motivated by uncontrollable female rage that cannot be deactivated and therefore ends up "a bloody mass," as typically occurs in war fantasies of violence visited upon women (Karl Theweleit, cited by Modleski, 1991, p. 62).

Not all film androids are so lethal or so feared as the uncontrollable Eve VIII. Humorously considering that a male robot might, like Frankenstein's monster, have a more sympathetic and humane character than its creator, Susan Seidelman's *Making Mr. Right* describes how a female public relations specialist is hired to teach social skills and graces to a male robot so that he might be made more endearing to the public. The private corporation

ChemTec developed Ulysses to do hazardous jobs that take a toll on humans, such as working with explosives and going into space for long periods. When government research funds for Ulysses might be withdrawn, the company hires Frankie Stone (Ann Magnuson) to promote their technology to the public, who can in turn exert influence on the distribution of government funds.

Despite her masculinized name, Frankie represents the feminine social world that must be brought into conjunction with the technological product promoted by ChemTec. She teaches the robot how to be human, learning in the process that she loves Ulysses because he is everything the uptight, obnoxious male scientist who created him is not. Although she and other single women represented in the film have had a difficult time finding the right men, Frankie quickly recognizes that Ulysses is more sincere, humane, and tender than his inventor Dr. Peters and the flesh-and-blood men who pursue her, the unctuous Dr. Ramdas and her on-again, off-again philandering politico boyfriend. While Frankie's attitudes advance the romance plot, her flaky feminine style and outrageousness are the inverse of the rational masculine attitudes framed by the male characters (with the exception of Ulysses), who are described as scientific and political authorities despite their personality flaws.

Chief Robotic Engineer Jeff Peters built Ulysses (both roles played by John Malkovich) in his own image, even giving the robot a penis so that he would have "confidence," an elusive characteristic for women in scientific and technical fields as social scientists document (Margolis & Fisher, 2002). Dr. Peters barely condescends to speak with Frankie, resenting attempts to humanize the mental, emotional, and social capacities of the android because he fears the robot designed to live independently will be spoiled by the socialization lessons. The scientist scoffs at Frankie's contributions and criticizes her for living in "an emotional swamp"—and for dragging Ulysses into that morass. That the android is a child open to new experiences is a conceit allowing Frankie to introduce him to society, including her friends. Ulysses reacts enthusiastically to all possibilities, especially when he manages to escape from the lab. He shops for a tuxedo with Frankie at a mall, dates the woman employee who has been chasing Dr. Peters, and manages to engage in a sexual encounter with Frankie's best friend Trish within one day. While all three women express pleasure at spending time with him, Trish's praise rings loudest: "He was so loving! So compassionate! So understanding of a woman's heart!" Even after she finds out he is an android, she is eager to repeat their sexual encounter.

In this film, the male scientist is the least likable character for his single-minded devotion to work. His refusal to recognize the value of human relationships and feelings make him a failure as a human being and a difficult employee and co-worker. Dr. Peters resents any human interaction as wasting time that should be devoted to science and working on his technological achievement; in contrast, Ulysses is kind, generous, and even romantic. In fact, creator and creation change places in the film because their characters reverse: the scientist becomes more robot-like and the robot becomes more human as the romance plot progresses. The male scientist's knowledge is represented in the film as narrowly conceived to achieve a place in the scientific pantheon, while Frankie's relationship with Ulysses delineates how the lack of technological expertise, combined with some knowledge of human relations, benefits society by enhancing romance. In sending Dr. Peters into space and permitting Ulysses to stay on earth with Frankie, the movie's plot resolves both Frankie's romantic problems and the robot's search for a significant human companion. The conclusion of Seidelman's film wryly represents technological innovations as having great consequences for humanity and the individual rather than expanding scientific and technical knowledge. *Making Mr. Right* suggests that if scientific experimentation and technological innovations could be directed toward the stereotypically feminine outcomes of helping others, the world might be a better place for most men and women. By sending the uptight Dr. Peters into space, the film suggests that the scientific initiative to find new knowledge has its limits on earth: while the scientist is a loner who eagerly undertakes a journey through the cosmos, society prefers humanized technological products to overly rational technoscientists.

Characterizations employing stereotypical attributes of men, women, robots, and technoscientists in both films construct related plots related to masculine fears about feminine power. Science and technology are inevitably overwhelmed by feminine desires: Eve's repression and Frankie's passion supersede their capacities to evaluate innovations. Male experts, especially those using scientific and technical tools and weapons, are delineated as more interested in vaunting technological expertise to gain professional prestige and general social approbation, while female experts are depicted as interested in technological pursuits for the sake of personal fulfillment and helping to improve the world. As feminist scholars of technoscience note, girls and women often express social interests as motivating their study and practice of science and technology and inspiring them "to transform the way in which science and technology is practiced" (Balsamo, 2000, p. 192).

Eve of Destruction and *Making Mr. Right* propose that distinguishing too carefully between human and machine is necessarily a failed endeavor, hinting that scientific interest in artificial intelligence and robotics is intertwined, perhaps in inappropriate ways, with meditating on the limits of being human. *Making Mr. Right* entertainingly represents what happens when a female publicist falls in love with an android who is destined to be sent into space, reconfiguring the traditional romance narrative as a story of a woman and a man-machine. While Seidelman's film suggests that robots can be better humans than their creators, *Eve of Destruction* describes what happens when a cyborg military weapon created by a woman for ostensibly peaceful purposes acts all too humanly in wreaking vengeance on the world.

Feminine values about science are represented as unconventional and somewhat threatening in these films: Dr. Simmons and Frankie recognize their own iconoclasm as their projects overturn traditional (i.e., masculine) approaches to robotics and artificial intelligence. Dr. Simmons is the only woman working on the military's robotics program; her work on the peacemaker cyborg resembles Frankie's attempt to humanize the android designed to work in similarly dangerous occupations. Although neither film definitively preaches that feminist values in scientific and technological enterprises will produce kinder, gentler (and therefore better) products, the feminist impulse submerged in both films suggests that public understanding and appreciation of science would increase if scientists and technologists worked to improve the human condition rather than to extend abstract knowledge or invent new technologies. The films discern different prospects regarding feminine influences on what are depicted as the masculine domains of science and technology, with *Eve*'s dark pessimism about the inevitable failure of the peacemaker android contrasting with *Mr. Right*'s optimism that cyborgs can engage in sympathetic human relationships.

Resisting feminine reconfiguration of scientific methods and technological outcomes, these films rely on common stereotypes about male and female attitudes toward love and work, suggesting that love and work are more significantly transformed by technoscience than transforming of it. Representing how cyborgs might break through gender stereotypes, showing the robot Eve as a better military operative than McQuade and Ulysses a better lover than Dr. Peters, the films suggest technology might transcend gender stereotypes (perhaps in destructive ways) even if humans cannot. *Eve of Destruction* represents a robot as a weapon tweaked by the female scientist to be a peace officer, but the robot's circuits revert to relying on stereotypically feminine motives (revenge because of abuse) in stereotypically masculine style (using guns to destroy enemies). *Making Mr. Right* outlines how feminine characteristics associated with nurturing cause a male android to become more human. Both films represent women as contributing feminine instinct and insight to a traditionally male purview in order to reinforce traditional roles assigned to men and women: *Eve of Destruction* criticizes the woman scientist whose repressed anger becomes her greatest personal contribution to her cyborg creation, while *Making Mr. Right* rewards the publicist for acting on her emotions and enabling the android to act on his. The films differ most on responding to the question of what science has accomplished in creating a robot as human as possible, representing our fears of the too-angry female cyborg and our desires for the loving male cyborg.

Both films update the Ovidian myth of Pygmalion and Galatea in which the inanimate female statue comes alive, for the scientist replicates him/herself in the shape of a robot. The artistic act of creating artificial life is represented as technological hubris in *Eve of Destruction* and *Making Mr. Right,* for both the male and female scientists engage with technology out of transgressive motivations, to fulfill personal desires rather than social needs. As Hillis Miller (1990) has observed, "Ovid's stories show that you always get some form of what you want, but you get it in ways that reveal what is illicit or grotesque in what you want" (p. 1). Transgressive characters in the *Metamorphoses* end up in a limbo state between life and death: "a memorial example still present within the human community. . . . a sign that his or her fault has not been completely punished or expiated" (p. 2). Galatea's transformation from inanimate to animate object reverses the typical transformation represented in Ovid's stories and is alluded to in contrasting ways by the robots Eve and Ulysses, for their animation signifies tragic destruction in the former and romantic fulfillment in the latter.

The plots of *Eve of Destruction* and *Making Mr. Right* represent conflicts between masculine and feminine views of technoscience without resolving them. By incorporating emotional, ethical, and other social concerns into their work, Dr. Simmons and Frankie Stone offer approaches to the scientific development of technology that differ substantially from their male colleagues. Personified by these female characters, science and its technological applications appear within the films as more sympathetic and humane in its outcomes than scientific ideas and technologies of earlier literature, more "female-friendly," to use Sue Rosser's (1997) phrase. But the efforts of Eve and Frankie to change the larger dimensions of scientific method and practice and to develop technologies incorporating feminist principles and addressing issues of social justice are not extensively treated in the films, which remain constrained in presenting the largely masculine and somewhat hostile environments of science and technology within the fantastic realm of science fiction. In an age when technologies such as computers and appliances with electronic chips are marketed specifically to women, and more women continue to study and practice in scientific and technological professions, viewers can hope that even the mildly feminist messages about technoscience suggested by *Eve of Destruction* and *Making Mr. Right* might inspire new cultural and film scripts representing a powerful conjunction of feminism and technology.

References

Balsamo, Anne. (1996). *Technologies of the gendered body: Reading cyborg women.* Durham, N.C.: Duke University Press.

———. (2000). Teaching in the belly of the beast: Feminism in the best of all places. In J. Marchessault & K. Sawchuk (Eds.), *Wild science: Reading feminism, medicine, and the media* (p. 185–213). New York: Routledge.

Baym, Nina. (2001). *American women of letters and the nineteenth-century sciences: Styles of affiliation.* New Brunswick, N. J.: Rutgers University Press.

Child, Lydia Maria. (1997). Hilda Silfverling. In Carolyn Karcher (Ed.), *The Lydia Maria Child reader.* Durham: Duke University Press. (Original work published 1845)

Cockburn, Cynthia, & Ormrod, Susan. (1993). *Gender and technology in the making.* London: Sage.

Colatrella, Carol. (2001). From *Desk Set* to *The Net:* Women and computing technology in Hollywood films, *Canadian Review of American Studies, 31*(2), 1–14.

Creager, Angela, Lunbeck, Elizabeth, & Schiebinger, Londa. (2001). *Feminism in twentieth-century science, technology, and medicine.* Chicago: University of Chicago Press.

Crichton, Michael. (1993). *Disclosure.* New York: Ballantine Books.

Ephron, Nora. (Director). (1993). *Sleepless in Seattle* [Motion picture]. United States: TriStar Pictures.

———. (Director). (1998). *You've got mail* [Motion picture]. United States: Warner Bros.

Etzkowitz, H., Kemelgor, C., & Uzzi, B. (2000). *Athena unbound: The advancement of women in science and technology.* Cambridge, U.K.: Cambridge University Press.

Faludi, S. (1991). *Backlash: The undeclared war against American women.* New York: Crown Publishers.

Fox, Mary Frank. (1998). Women in science and engineering: Theory, practice, and policy in programs. *Signs, 24*(1), 201–223.

Fox, Mary Frank. (1999). Gender, hierarchy, and science. In J. S. Chafetz (Ed.), *Hand-*

book of the sociology of gender (pp. 441–457). New York: Kluwer Academic/Plenum Publishers.

————. (2001). Women, science, and academia. *Gender & Society* 15 (5), 654–666.

Gibbins, Duncan. (Director). (1991). *Eve of Destruction* [Motion Picture]. United States: Interscope Communications.

Gibson, William. (1984). *Neuromancer.* New York: Berkley/Ace Books.

Haraway, Donna. (1998). A cyborg manifesto: Science, technology, and socialist-feminism in the late twentieth century. In P. Hopkins (Ed.), *Sex/Machine: Readings in culture, gender, and technology* (pp. 434–467). Bloomington: Indiana University Press. (Original work published 1985, 1991)

Harding, Sandra. (2001). After absolute neutrality: Expanding "science." In M. Mayberry, B. Subramaniam, & L. H. Weasel (Eds.), *Feminist Science Studies* (pp. 291–304). New York: Routledge.

Hawthorne, Nathaniel. (1987). *Selected tales and sketches.* New York: Penguin.

Hopkins, Patrick (Ed.). (1998). *Sex/Machine: Readings in culture, gender, and technology.* Bloomington: Indiana University Press.

Khouri, Callie (Writer). Scott, Ridley (Director). (1991). *Thelma and Louise* [Motion picture]. United States: Metro-Goldwyn-Mayer.

Kristeva, Julia. (1981). Oscillation du "pouvoir" au "refus" [Oscillation between power and denial]. In Elaine Marks & Isabelle de Courtivron (Eds. & Trans.), *New French feminisms: An anthology* (pp.164–166). New York: Schocken Books.

Lang, Walter. (Director). (1957). *Desk set* [Motion picture]. United States: Twentieth Century Fox.

Levinson, Barry. (Director). (1994). *Disclosure* [Motion Picture]. United States: Warner Bros.

MacKenzie, Donald, & Wajcman, Judy. (1999). Preface to the second edition. In D. MacKenzie and J. Wajcman (Eds.), *The social shaping of technology* (2nd ed.) (pp. xiv–xvii). Buckingham, U.K.: Open University Press.

Margolis, Jane, & Fisher, Allan. (2002). *Unlocking the clubhouse: Women in computing.* Cambridge, Mass.: MIT Press.

Martin, Steve (Writer). Schepisi, Fred. (Director). (1987). *Roxanne* [Motion picture]. United States: Columbia Pictures.

McIlwee, Judith S., & Robinson, J. Gregg. (1992). *Women in engineering: Gender, power, and workplace culture.* Albany: State University of New York Press.

Merchant, Carolyn. (1990). *The death of nature: Women, ecology, and the scientific revolution.* San Francisco: Harper.

Miller, Hillis. (1990). *Versions of Pygmalion.* Cambridge, Mass.: Harvard University Press.

Modleski, Tania. (1991). *Feminism without women: Culture and criticism in the postfeminist age.* New York: Routledge.

Noble, David. (1992). *A world without women: The Christian clerical culture of Western science.* New York: Knopf.

Oblepias-Ramos, Lilia. (1998). Does technology work for women too? *Sex/Machine:*

Readings in culture, gender, and technology. In P. Hopkins (Ed.), *Sex/Machine: Readings in culture, gender, and technology* (pp. 89–94). Bloomington: Indiana University Press. (Original work published 1991)

Oldenziel, Ruth. (2001). Man the maker, woman the consumer: The consumption junction revisited. In N. Creager, E. Lunbeck, & L. Schiebinger, *Feminism in Twentieth-Century Science, Technology, and Medicine* (128–148). Chicago: University of Chicago Press.

Piercy, Marge. (1993). *He, she, and it* (2nd ed.). New York: Fawcett Crest/Ballantine. (Original work published 1991)

Porush, David. (1992). Transcendence at the interface: The architecture of cyborg utopia, or cyberspace utopoids as postmodern cargo cult. American Library Association: available at http://www.cni.org/pub/LITA/Think/Porush.html.

Pursell, Carroll. (2001). Feminism and the rethinking of the history of technology. In Angela Creager, Elizabeth Lunbeck, & Londa Schiebinger (Eds.), *Feminism in twentieth-century science, technology, and medicine* (pp. 113–127). Chicago: University of Chicago Press.

Rosser, Sue V. (1997). *Re-engineering female friendly science.* New York: Teachers College Press.

Schepisi, Fred. (Director). (1994). *IQ* [Motion picture]. United States: Paramount Pictures.

Scott, Ridley. (Director). (1982). *Blade runner* [Motion picture]. United States: Embassy Pictures.

Shelley, Mary. (1991). *Frankenstein, or the modern Promethus* (M. K. Joseph, Ed.). Oxford, U.K.: Oxford University Press. (Original work published 1818)

Seidelman, Susan. (Director). (1987). *Making Mr. Right* [Motion picture]. United States: Orion Pictures Corporation.

Sonnenfeld, Barry. (Director). (1997). *Men in black* [Motion picture]. United States: Amblin/Columbia.

Stanley, Autumn. (1998). Women hold up two-thirds of the sky. In P. Hopkins (Ed.), *Sex/Machine: Readings in culture, gender, and technology* (pp. 17–32). Bloomington: Indiana University Press. (Original work published 1983)

Villiers de L'Isle-Adam, Auguste. (1982). *L'Eve future [Tomorrow's eve].* (Robert Martin Adams, Trans.). Urbana: University of Illinois Press. (Original work published 1886)

Wajcman, Judy. (1991). *Feminism confronts technology.* University Park, Pa.: Pennsylvania State University Press.

Warner, Marina. (1994). *From the beast to the blonde: On fairy tales and their tellers.* New York: Noonday Press/Farrar, Straus and Giroux.

Winkler, Irwin. (Director). (1995). *The net* [Motion picture]. United States: Sony Pictures.

Zemeckis, Robert. (Director). (1997). *Contact* [Motion picture]. United States: Warner Brothers.

9

High-Tech Worship: Gender Politics and the Appropriation of Multimedia Technology for Christian Worship

JAMES FENIMORE

Over the past decade many Protestant Christian churches in the United States have incorporated multimedia technology into their worship practice. As an ordained United Methodist pastor, I have been involved with the development and implementation of media ministry over the past decade. The assumption that I (and those involved in media ministry) have made is that the technologies incorporated are simply tools. This chapter will trace the adaptation and appropriation of these wide-ranging technologies in the worship setting and explore the question, "Do multimedia technologies have gender politics?" A content and discourse analysis of the widely read magazine *Technologies for Worship*,[1] a magazine devoted to the appropriation and adaptation of multimedia technology for the church, provides evidence of gender exclusion in print advertisements and editorial content and establishes a connection between the gender politics of "evangelical" churches and the media ministry that traces its roots to these churches. Exploring the categories of the sacred and the profane (Douglas, 1996; Durkheim, 1995) reveals how technological artifacts have been made sacred while at the same time women have continued to be excluded as profane.

In the past decade, many Christian churches of various denominations have converted their worship spaces into multimedia screening rooms. It is estimated that as many as 50 percent of all churches are conducting a media worship service or have plans to start one.[2] Christian worship, which has relied primarily on oral communication, sacred texts, and sacred spaces since its inception, is now being assisted by multimedia "experiences" geared to attract younger members. Cultural relevance and an almost desperate need for new members (and their financial resources) has led to a wholesale acceptance of almost any means that might draw the elusive "younger generation" back to the church once again. A graying congregation, combined with a cumbersome and costly institutional structure, has led to a search for the magic elixir that will revive the church. Many consider media ministry[3] this elixir. Wholesale acceptance of multimedia technology carries with it various political and gender implications.

The success of churches like Ginghamsburg United Methodist Church (Tipp City, Ohio), a pioneer in media ministry, has led to the replication of their worship style and use of media. Popular Protestant theologians (Riddell, 1998; Sample, 1998; Slaughter, 1998; Sweet, 1999) have offered theological justifications for the adoption of technology; using the analogy of the adoption of the printing press as the fuel of the Protestant Reformation, they claim this media reformation will be as revolutionary.

I have led numerous workshops and lectures on the importance of multimedia technology in the church; I (Fenimore, 2001) and others (Barbour, 1993; White, 1994) have written about technology as if it were a value-neutral tool to enhance worship. However, using the theories of Science and Technology Studies (STS) and the method of content analysis reveals that the multimedia technologies churches are using are far from value-neutral. In fact, the gender politics of these technologies are in direct conflict with the stated social principles of my own denomination (which takes pride in its inclusive political stance) and incarnate the gender stereotypes and divisions held by many (if not most) Protestant Christian churches.

The focus on gender in my research, as opposed to class or race, has significance. I would suggest the foundational issue that separates conservative and liberal churches is the issue of gender roles. These roles embodied in church or denominational polities describe clearly the limitations, if any, placed on women. As the "media ministry" worship style begins to replicate itself and move beyond its conservative roots, I can't help but wonder, "Is there something more than artifacts and techniques being transferred?"

To study this I suggest employing the use of a "hermeneutic of omission"[4] that implements a radical philosophical reorienting. Instead on focusing on an interpretation of what is *seen* (usually in texts, but I expand this to images as well), the focus of the hermeneutic of omission is on what is *unseen*. The unseen can provide an insight into the politics of the seen. Raising questions about how often women are depicted (or not depicted) in the media ministry advertisements and discourse is an important start to this process. In the end, being suspicious of these omissions will only help to make media ministry a more inclusive representation of the church it seeks to re-present.

Soul Café: A Media Worship Experience

It's a Sunday evening, and people are gathering for worship at my church in downtown Troy, New York. They enter not through the ornate sanctuary doorways, but through a side door that leads them to a hall, which at first glance seems like any church hall. Details within the room begin to become apparent. On the stage is a large screen lit with a picture and the words, "Welcome to Soul Café." The room is filled with small, round tables with lighted candles on them. A table in the back of the room discreetly houses numerous technological devices that give off an electronic hum. A person walks up to the front of the room to a stool and microphone to offer a welcome and introduce a musician.

The musician is a flamenco guitarist who lives in Troy. She is not a member of the church, nor is her music "Christian" by definition. But as she plays, everyone in the room sees into the soul of an artist and is mesmerized by the power of her music. On the screen is a photograph of her recently released CD. After she has finished several songs, a pastor gets up and tells a personal story about a painful episode in her life. She ends by introducing the theme for the evening, which is entitled, "I Need My Pain." A movie clip is introduced; it is a clip from *Star Trek V*, in which Captain Kirk explains that he doesn't want the pain in his life magically taken away; he needs it because it is a part of who he is. Another pastor gets up and talks about pain in our lives and how, as much as we wish we might be able to take some magic pill and relieve ourselves from it, we cannot. He relates the story told in the movie to the sacred Biblical story, describing how the early Christians suffered and were persecuted for their beliefs, and describing the powerful witnesses of those individuals whose writings we read in the Bible. The flamenco guitarist is introduced again and plays several more songs, teaching about the meaning and the passion behind each piece of music. At the end of the program, the pastor poses the question, "Do you need your pain?"

This service, which runs as smoothly as a television production, is made possible through a network of relatively inexpensive technical equipment and technicians (who are usually church volunteers) who have not been trained in multimedia technology; their expertise has come from "doing

Figure 9.1. Schematic of the Technologies Used in Soul Café

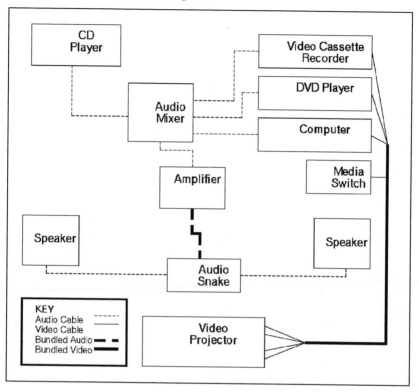

multimedia." The construction of the controlling technologies of this service occurred over the past three years through a tinkering process. In Figure 9.1, a schematic of the technologies is illustrated. The two primary technologies are the projector (through which video images are processed) and the audio mixer (through which all audio signals are processed). Through the controls of these technologies, a technician can move from VHS clip to DVD clip to computer graphics seamlessly. It is a great example of the adaptation of multimedia technology, intended for home use, redesigned for use in a religious setting.

The Adaptation/Appropriation of Multimedia Technology

The words adaptation and appropriation are sometimes used interchangeably, but they are not the same. Ron Eglash (2004) describes these terms as existing along a continuum based on consumption and production. The movement from consumer to producer parallels the level of appropriation. Eglash begins on the consumer end of the spectrum with "reinterpretation," which only changes the language associated with an artifact. Next and somewhat in the middle of this spectrum is "adaptation" which modifies the artifact by using it in a different way than was originally conceived. The greatest changes are "appropriation" where the artifact is structurally changed; this lies on the production end of the spectrum.

The use of technology in media worship is not just at the level of adaptation of technologies; there are also examples of appropriation of technologies. Microsoft's PowerPoint has long dominated the presentation-software market. But PowerPoint, designed to move from slide to slide in a linear fashion, has limitations and is difficult to use in the setting of religious worship service. In other words, an assumption of the PowerPoint design is that the user will move from slide 1 to slide 2, etc. This may be appropriate for a business presentation, but can be an awkward restriction to a free flowing worship service. A company formed by media ministry technicians,[5] Grassroots Software, developed their own presentation software to solve these problems. A common task in media ministry is the projection on the screen of lyrics to a song that the congregation is singing. One of the problems this software solves is the need for the technicians to be able to move quickly from one song's lyrics to another without prior knowledge of the order of songs. Grassroots Software accomplishes this by having a database of preselected songs available *in the program*, ready for use.

The theology of the preselected songs is a very conservative representation of the Christian repertoire. This is no accident. The core developers of this software all were employed by Promise Keepers, a men's spirituality ministry and theologically/politically conservative organization founded by former football coach Bill McCartney in 1990. Its mission (according to its website) is for "igniting and uniting men to be passionate followers of Jesus Christ through effective communication of the 7 Promises."[6]

Another company founded by Christian technologists is Lumicon Digital Productions. Funded by UMR Communications (a company related to the United Methodists), Lumicon provides complete multimedia packages for worship services. The packages include graphic slides (ready to be used in any presentation software package) for use during services, and original animations and video clips that all revolve around a central theme that Lumicon has developed. An example of one of these themes is "Good to the last drop," which they describe in more detail as allowing God "to filter the good in from the bad." The team provides all of the media resources needed to conduct a multimedia worship service (assuming that you have the hardware and software needed to run the media).

The core team of Lumicon technologists includes two major figures in the field, Len Wilson and Jason Moore. Both came from the Ginghamsburg United Methodist Church, where they helped launch one of the first and easily the most successful media ministries to date. Their productions are complete worship experiences, professionally designed by talented graphic artists and designers. A church that does not have members with technical expertise can nevertheless purchase and produce these kinds of services; the down side is that the creative process is handed over to people outside of the community in which these productions will be used. This can lead to the politics of the community and the artifact being at odds.

The alternative is for the development of expertise in the local setting. A magazine devoted to this mission is *Technologies for Worship;* published in Canada, it has a wide distribution throughout the United States. By the end of 1999, their circulation was 5,000; in just over a year, it had mushroomed to over 30,000. Their only rival is a magazine that is distributed for free (paid by its advertisers) to about 1,500 people.

I began reading this magazine ten years ago, intending to enhance my technical skills. I found messages in the advertisements and the feature articles that were a strange mix of religious zeal and glorification of the technological artifact (see Figures 9.2 and 9.3). I found myself asking the question, "If the companies who produce these products have a particular ideological and theological stance, then do the artifacts they produce have the same politics?"

The Gender Politics of Conservative Evangelicals

In order to understand the role of gender in multimedia technologies in the church, it is important to review the gender politics of conservative evangelicals. The appropriation of multimedia technology began with conservative evangelicals, and many of the producers of multimedia technologies and their content have a connection with evangelical Christians. Historically, the term "evangelical" has been applied to all Christians who believe in the Gospels of the New Testament. More recently, the descriptor "evangelical" has become associated with conservative Christians who have opposed the militancy of fundamentalists (Harding, 2000).[7] Similar to fundamentalists, evangelicals are often Biblical literalists and politically conservative. Unlike fundamentalists, evangelicals are not separatists. They engage the world and are quick to use the technologies and methods associated with the world.

Evangelicals have a patriarchal worldview that is both prescribed and confirmed by their interpretation of the Bible. The initial creation of man and the subsequent creation of woman as man's helpmate[8] define the hierarchical role of man over woman. This "created order" defines the household roles of men and women. Men are therefore the head of the household. Other Biblical texts[9] reinforce this "created order" and establish the man as the "spiritual head of the household."

These interpretations of the Bible have led to the exclusion of women as spiritual leaders. Clergy status in conservative evangelical churches is reserved for men.[10] Women are given some leadership roles within the evangelical church, including children's ministries and music. The authoritarian hierarchy with a male pastor at the top of this structure ensures that, regardless of the contributions of women in leadership roles, the male pastor has control of the ultimate decision-making authority.

Sacred and Profane

At the root of evangelical gender politics is an understanding of a distinction between the sacred and the profane. Emile Durkheim (1995/1912) has produced the foundational work on these categories in his text *The Elementary Forms of Religious Life*. Durkheim stressed that there needed to be a distinct separation between that which is considered sacred and that which is considered profane. These categories could not mix without defiling the sacred. Mary Douglas (1966) in her text *Purity and Danger* expands on this notion: "To talk about a confused blending of the Sacred and the Unclean is outright nonsense" (Douglas, 1966).

These boundaries of the sacred and profane have been simultaneously removed and upheld. They have been removed, or (in the language of Mary Douglas) there has been a "confused blending," in terms of materialism. Artifacts that have been constructed as profane tools of business have been elevated to the status of the sacred by the companies that produce them and the churches that appropriate them.

I found that many of the advertisements in *Technologies for Worship* were explicitly co-opting theological language in an attempt to illustrate how the technology fits into a religious setting. Figure 9.2 is an advertisement of the newest version of Grassroots' presentation software. The use of the phrase "Version Birth" is a blending of the sacred doctrine of the virgin birth and the profane product of a new version of software. The producers of this software were hesitant to use this language in the advertisement, worried that readers might be offended, but they decided to go ahead with the ad. Not a single complaint was voiced.[11]

Another example, from Audio-Technica selling wireless microphones, appears in Figure 9.3. The image is of a woman using an Audio-Technica microphone in what appears to be a worship service. The text that stands out on the ad is "Make a Joyful Noise," which happens to be the beginning of Psalm 100, which reads "Make a joyful noise unto the Lord." The use of sacred text to sell a product is becoming a common marketing strategy. Sometimes this is accomplished simply through the language used, other times it is a more subtle use of images. In the end, the result is the same: a "confused blending" of the sacred and profane—and it is not confined to multimedia technology.

Colleen McDannell (1995) in her text *Material Christianity* illustrates a wide range of this "blending" in the consumer products of Christianity. Studying the material culture of Christians in the United States in the late twentieth century, McDannell focuses her research not on what is said but what is done (or in this case what is purchased). From the Family Bible to "Christian Kitsch," McDannell provides numerous examples of the eradication of the separation of sacred and profane.

McDannell finds fault with the exclusive categories of sacred and profane. She notes that there are foundational doctrines in Christianity that subvert the exclusiveness of these categories. The doctrine of the incarnation is especially relevant to our discussion. In this doctrine, God (sacred) becomes human (profane) in the person of Jesus. The "blending" of what should be two mutually exclusive categories, sacred and profane, produces the most powerful expression of the Christian faith, God incarnate (McDannell, 1995; Fenimore, 2001).

The elevation of the profane to sacred is central to media ministry. I collected statements on the use of technology in worship from the "Editor's Notes" section of each of the magazines (Table 9.1). Multimedia technologies are described as: modern tools, a means of effective communication, a weapon of a modern day religious soldier, controlling the way we think, a way to communicate the message to a younger generation, God-given tools, and a way to communicate to a broader audience.

Using these statements, one gains a portrait of media ministry and the role of technology in the church. Technology is seen as a God-given tool that can be used to communicate the message of the church in a more effective way.

Figure 9.2. Ad for Grassroots Software in *Technologies for Worship*

Figure 9.3. Ad for Audio-Technica Microphone in *Technologies for Worship*

Table 9.1: Statements on Worship and Technology from *Technologies for Worship*

An era where the church will have a stronger voice by using all the modern tools available today
The challenge is to ensure that your church or ministry stays on top of how to effectively communicate the message to reach your people.
We are all modern-day soldiers—fighting to keep God's message alive and well in a world driven by high technology.
Multimedia is controlling how we think, relate, and learn, and all the while, these technologies are starting to amalgamate together.
Technology is moving faster than most of us realize, and with that in mind, it is great to know that churches are adapting to these breakthroughs to communicate to younger generations.
How exciting to think that we have this opportunity to use the tools God has given us to portray the most important message of all.
Implementing new technologies to augment worship services is helping to communicate the Message to a broader audience of people.

Nothing about these technologies' use outside of the church is considered in these statements.

The multimedia technologies used by churches are considered "high tech." They are expensive technologies to acquire and have a high cost to maintain. These technologies are also highly valued by our society and represent the building blocks of our "technological society." This high value, both financial and cultural, may also be a clue to the ease with which this technology is appropriated into the church.

Women and Media Ministry

The elevation of artifacts to sacred status has not helped break down the boundary between men and women. Women viewed as "the other" in this patriarchal society have not been elevated to the status of sacred. This exclusion can be quantified.

I began to simply count the number of women and men portrayed in the ads of *Technologies for Worship* (see results in Table 9.2). These ads depicted primarily technological artifacts, but individuals were also displayed. There were more males depicted than females, as can be seen in these results.

Looking even closer at the men and women depicted in the advertisements, we can categorize the role of each actor as either a User of the technology, a Controller of the technology, or as having No Role. A "No Role" designation, the most frequent, was assigned to ads with pictures of faces or entire bodies that had no relation to the technology and represented exactly 77 percent of

Table 9.2: Counts of Ads in *Technologies for Worship* (*TFW*)

Issue of *TFW*	Males in ads	Females in ads	Non-whites in ads*	Total number of ads
January 2000	5	1	0	38
March 2000	3	2	0	43
May 2000	7	2	0	43
July 2000	12	7	2	60
Sept. 2000	3	2	0	54
January 2001	7	6	3	40
March 2001	18	5	2	52
May 2001	10	12	2	52
July 2001	14	16	3	74
Sept. 2001	4	3	0	54
Nov. 2001	4	2	0	48
Totals	87	58	12	558

* The count of how many non-whites are depicted in the advertisements of this magazine was an addition to my original research goals. I place the results in this table to illustrate there is a great need to study race and technology in media worship.

both the women and the men depicted in these ads. Excluding this category and looking simply at Users versus Controllers of the technology, women are overwhelmingly depicted as Users (69 percent of the time) and men are overwhelmingly depicted as Controllers (62 percent of the time).

Of greater interest in terms of gender politics was a simple count of the number of women authors who contributed to the content of the magazine. Table 9.3 illustrates the results of this tally. The stark contrast of 173 male authors to 19 female authors is even greater than these numbers suggest when we remove the couples who coauthored articles; then we are left with 161 males and just 7 females.

Comparing these results with a gender count of authors of the similar but nonreligious magazine *Presentations,* we find that there is a notable difference between the two. During the same period,[12] 59 percent of all articles in *Presentations* were authored by males, 41 percent by females. This is compared with the results of *Technologies for Worship* where 90 percent were authored by males and 10 percent by females. The difference in these two percentages points to a quantifiable exclusion of women from leadership roles in media ministry.

The lack of women in this newly emerging field has been explicitly addressed by contributors to the magazine. In the January 2001 issue of *Technologies for Worship,* a column that highlights some of the discussion happening on the electronic bulletin board included the following comment by "audiowoman" entitled "Men vs. Women":

Table 9.3: Counts of Authors' Gender in *Technologies for Worship*

Issue of *TFW*	Male authors	Female authors
January 2000	16	2
March 2000	19	2
May 2000	15	2
July 2000	14	1
September 2000	13	2
January 2001	15	1
March 2001	16	4
May 2001	19	2
July 2001	16	0
September 2001	17	2
November 2001	13	1
Totals	173	19

I'm an up-and-coming audio video tech and have begun working with our multimedia ministry, which is ALL MEN. I always wanted a career in sound and have pursued some classes but need more "hands on" training. Why do men think that women cannot be technically inclined, especially in this field? The guys I work with totally ignore me when we are working in the video room. And these are supposed to be born-again Christian guys? Can any male out there shed any light or information on what I can do? Is there any female out there who has gone through the same thing?

—audiowoman

A response to "audiowoman's" concerns by "Steve" (reprinted with typographical errors corrected):

I hope the title Men VERSUS women is not an indication of how you see the issues with regard to the techie side of Church life. It shouldn't be a competition . . . but unfortunately it often turns out that way even between men. Bear in mind that we are human and that we do need the Lord's help. I would love some of the females of our church to take an interest in the production side of things. I must admit to sometimes feeling out of place, particularly when doing the sound for Womens [*sic*] events. Keep plugging away there and show them men how it should be done.

—Regards, Steve

What is lacking in this and the other responses[13] to "audiowoman's" concern is any understanding there is something deeper going on here. There is never even a hint that systemic sexism is the problem. In fact "Steve" says he would welcome women in this role if they just would "take an interest" in this field. The problem of women and technology seems to be reduced to an issue of women not being interested in the technology, and "if only they would take an interest" in this, everything would be great. But "audiowoman" is interested; in fact, she is quite skilled—and yet she feels rejected, excluded.

Systemic sexism is alive and well in the conservative evangelical churches. These churches, which are often independent or unaffiliated with any denomination, have strict polities that exclude women from leadership positions, especially the role of pastor. Mainline churches, defined by their longstanding denominational affiliations, have polities that are inclusive of women in the leadership structure. Women constitute nearly 35 percent of all seminary students training to be clergy.[14] My own denomination, the United Methodist Church, has ordained women as pastors since 1956, and women have attained all levels of leadership, including the highest level, bishop. Evangelical churches ascribe to the natural order, not inclusivity. The natural order is defined by the "created order" that prescribes "man's dominion over all of creation."[15]

Nature and Artifacts

Sherry Ortner (1974) suggests women are devalued because of the assumption that there is a hierarchy of culture over nature. Culture represents that which is controlled; nature represents that which cannot be controlled, but that which we rely on. Ortner suggests that women are symbolized as closer to nature and therefore they are devalued. Marjorie Procter-Smith, a feminist liturgical[16] scholar, also points out that women symbolically represent nature and the natural (Ruether, 1983; Procter-Smith, 1990). Technology lies outside the realm of nature, as it is controllable to those who have the power to master it. Technological artifacts represent more than simply "things"; they represent "the physical and mental know-how to make use of those things" (Wajcman, 1995, p. 201). This dichotomy between nature and artifacts describes the core division between men and women.

The hypermasculine reaction of feminine exclusion in the patriarchal institution of evangelical churches is at its core a perpetuation of the boundary between sacred and profane. Technological artifacts are equated with the sacred. This is accomplished by their important role of communicating the Gospel message. As instruments of the sacred act of "evangelizing," these artifacts take on a sacredness of their own. At the same time, women and their association with nature are relegated to the profane.

The upholding of this dichotomy of nature and artifact is the power behind the exclusionary treatment of women in media ministry. The taboo of combining these two categories is an unspoken affirmation of the systemic sexism inherent in the evangelical churches and consequently media ministry.

Conclusion

It is time to return to our overarching question: "Do multimedia technologies have gender politics?" If we are asking if a video projector is sexist, or contains within it sexism, that is hard to say. If we are asking if the designers and engineers of these technologies have intentionally designed artifacts to exclude women and promote others, I say that I cannot accept this conclusion. But if the question is whether it is probable that the artifacts constructed will affirm the systemic sexism within our society, then clearly the answer is yes. The artifacts of media ministry may not be sexist, but they are currently being designed and used in ways that perpetuate sexism.

We have seen how gender and technology are intertwined in a complex and extremely subtle process that results in an artifact that represents the patriarchal culture of its designers. Although the artifact itself is not sexist, its design and implementation does reinforce the social order of its creators. In this chapter we have seen how the appropriation of multimedia technologies by conservative evangelicals has reinforced the "created order" and upheld the limited role women have in these churches.

The elevation of technological artifacts to the status of sacred is a powerful example of the value placed on technology by conservative evangelicals. Choosing to use technology to communicate the Gospel message *before allowing women to do so* signifies that conservative evangelicals would rather seek their "salvation" in technology. Framing this patriarchal worldview in the symbolic language of religion, conservative evangelicals have established a powerful weapon to perpetuate sexism.

Notes

1. A sample of two years (January 2000—December 2001, excluding one unavailable issue) were reviewed.

2. From Lumicon Digital Productions Web site, http://www.lumicon.org.

3. The term "media ministry" is an odd term that has developed to describe this form of worship. It is used by those who practice media ministry.

4. The hermeneutic of omission is a methodological tool for re/viewing texts or artifacts to seek what or who is missing. Building on the concept of the hermeneutic

of suspicion (Palmer, 1969; Ricoeur, 1974; Gadamer, 1975), this method uses a critical eye (or suspicious eye) to review exclusions in texts and artifacts. This method is most helpful in the process of reviewing movie clips, scripts, and graphical images in multimedia churches.

5. A new job title in these churches is "media minister," a technologist employed to construct media worship.

6. The Seven Promises of a Promise Keeper found on www.PromiseKeepers.org are:

1. A Promise Keeper is committed to honoring Jesus Christ through worship, prayer and obedience to God's Word in the power of the Holy Spirit.

2. A Promise Keeper is committed to pursuing vital relationships with a few other men, understanding that he needs brothers to help him keep his promises.

3. A Promise Keeper is committed to practicing spiritual, moral, ethical, and sexual purity.

4. A Promise Keeper is committed to building strong marriages and families through love, protection and biblical values.

5. A Promise Keeper is committed to supporting the mission of his church by honoring and praying for his pastor, and by actively giving his time and resources.

6. A Promise Keeper is committed to reaching beyond any racial and denominational barriers to demonstrate the power of biblical unity.

7. A Promise Keeper is committed to influencing his world, being obedient to the Great Commandment (see Mark 12:30–31) and the Great Commission (see Matthew 28:19–20).

7. The best description of the fundamentalist movement is Susan Friend Harding's, *The Book of Jerry Falwell.* Harding traces the alliance of fundamentalists and conservative evangelicals into "born-again Christians" who formed Falwell's Moral Majority and broke fundamentalist taboos by engaging in political activism and moving away from the separatist origins of fundamentalism.

8. Genesis 2.

9. 1 Corinthians 7, 11, 14; Ephesians 5; Colossians 3; 1 Peter 3.

10. There are examples of female evangelical preachers, but these churches are considered more liberal because of their lack of literal interpretation of the scripture.

11. This was communicated to me by the marketing and sales director of Grassroots Software during a phone interview on June 24, 2002.

12. The years 2000–2001.

13. The magazine's publication of "audiowoman's" concern had four responses on the electronic bulletin board, two by men and two by women. The editors of the magazine chose only to print the men's responses ("audiowoman" did, however, solicit the comments of men at the end of her note).

14. According to statistics collected by the Association of Theological Schools in the United States and Canada. These figures are for 2000 and represent the vast number of seminaries through North America. Clergy of independent and nondenominational churches often are not required to have any seminary training.

15. This comes from Genesis 2.

16. This refers to the study of worship.

References

Barbour, I. (1993). *Ethics in an age of technology.* San Francisco: Harper San Francisco.

Douglas, M. (1966). *Purity and danger: An analysis of concepts of pollution and taboo.* London: Routledge.

Durkheim, E. (1995). *The elementary forms of religious life.* New York: The Free Press.

Eglash, R., J. Crossiant, Di Chiro, & D., Fouché (Eds). (2004). *Appropriating technology: Vernacular science and social power.* Minneapolis: University of Minnesota Press.

Fenimore, J. (2001). *How a congregation's identity is affected by the introduction of technology-based ministries.* Unpublished doctoral dissertation, Drew University, Madison, N.J.

Gadamer, H. G. (1975). *Truth and method.* New York: Seabury Press.

Harding, S. F. (2000). *The book of Jerry Falwell: Fundamentalist language and politics.* Princeton, N.J.: Princeton University Press.

McDannell, C. (1995). *Material Christianity: Religion and popular culture in America.* New Haven, Conn.: Yale University Press.

Ortner, S. B. (1974). Is female to male as nature is to culture? In M. Z. Rosaldo & L. Lamphere (Eds.), *Woman, culture and society* (p. 67–87). Palo Alto, Calif.: Stanford University Press.

Palmer, R. E. (1969). *Hermeneutics: Interpretation theory in Schlieiermacher, Dilthey, Heidegger, and Gadamer.* Evanston, Ind.: Northwestern University Press.

Procter-Smith, M. (1990). *In her own rite: Constructing feminist liturgical tradition.* Nashville, Tenn.: Abingdon Press.

Ricoeur, P. (1974). *The conflict of interpretations: Essays in hermeneutics.* Evanston, Ind.: Northwestern University Press.

Riddell, M. (1998). *Threshold of the future.* London: SPCK.

Ruether, R. R. (1983). *Sexism and God-talk: Toward a feminist theology.* Boston: Beacon Press.

Sample, T. (1998). *The spectacle of worship in a wired world: Electronic culture and the gathered people of God.* Nashville, Tenn.: Abingdon Press.

Slaughter, M. (1998). *Out on the edge: A wake-up call for church leaders on the edge of the media reformation.* Nashville, Tenn.: Abingdon Press.

Sweet, L. (1999). *Soul tsunami: Sink or swim in the new millennium culture.* Grand Rapids, Mich.: Zondervan Publishing House.

Wajcman, J. (1995). Feminist theories of technology. In G. M. Shelia Jasanoff, James Petersen, & Trevor Pinch, *Handbook of science and technology studies* (189–204). London: Sage Publications.

White, S. J. (1994). *Christian worship and technological change.* Nashville, Tenn.: Abingdon Press.

Postscript:
Join the Conversation

MARY FRANK FOX, DEBORAH G. JOHNSON,
AND SUE V. ROSSER

This book initiates what we anticipate will be a series of books on intellectually stimulating, challenging, and ever-expanding areas of theory and research on women, gender, and technology. As Deborah Johnson says in the introduction, a primary impetus for the series arose from the dearth of materials on women, gender, and technology available to support undergraduate and graduate teaching and research. Later volumes in this series will explore particular technologies such as computers or particular aspects of gender, such as social networks. However, we chose consciously to make this first volume a relatively broad anthology. This choice came from our desire to portray the potential range and variety of issues and approaches in the study of women, gender, and technology.

As the chapters by Barbara Katz Rothman and Linda Layne signal, reproductive technologies form one strong focal point, drawing significant attention; information technologies (IT), addressed in the chapters by Cheryl Leggon, Judy Wacjman, James Fenimore, and Sue Rosser, serve as another magnet of interest. Although these new technologies fascinate us all, as we seek to understand their co-creation with gender, considerable room remains for examination of the interaction among women, gender, and the more traditional technologies, such as household appliances, cars, and buildings.

Mary Frank Fox's chapter on gender and status among engineers with doctoral-level degrees and Mara Wasburn and Susan Miller's chapter on technical education exemplify training, workforce, and organizational factors that affect who becomes and remains a designer, maker, and transmitter of technology and its meanings. Cheryl Leggon, Sue Rosser, Barbara Katz Rothman, and Judy Wacjman outline various factors that intersect with gender, including race, nationality, and class, and that differentiate users of technology and impact upon design of technologies. Different media such as literature and film, as explored by Carol Colatrella in her chapter, or the use of new technologically sophisticated multimedia in the traditional environment of religion and the church, as James Fenimore examines in his chapter, provide fresh perspectives on the co-creation of gender and technology.

Although this book is wide ranging in its topics and diverse in the disciplinary backgrounds and approaches of its authors, we recognize that it just begins to portray the many-faceted relationships between gender and technology among cultures, institutions, and individuals. We extend this inaugural volume as an invitation for others to continue with contributions to the understanding of women, gender, and technology through publication in this new series.

Contributors

CAROL COLATRELLA is a professor of literature and cultural studies in the School of Literature, Communication, and Culture, and the codirector of the Center for the Study of Women, Science, and Technology at Georgia Institute of Technology. In addition to publishing articles analyzing nineteenth- and twentieth-century American and European literary, historical, and scientific narratives, she has coedited an anthology, *Cohesion and Dissent in America*, and written two books, *Evolution, Sacrifice, and Narrative: Balzac, Zola, and Faulkner* and *Literature and Moral Reform: Melville and the Discipline of Reading*. She is currently working on a book to be titled *Toys and Tools in Pink: Cultural Narratives of Gender, Science, and Technology*.

JAMES FENIMORE is the senior pastor of Christ Church in Troy, New York and is a Ph.D. candidate at Rensselaer Polytechnic Institute in the science and technology studies department. His research focuses on the use and appropriation of media technology in Christian worship.

MARY FRANK FOX is NSF Advance Professor in the School of Public Policy, and the codirector of the Center for the Study of Women, Science, and Technology at Georgia Institute of Technology. Her research, focusing upon women and men in academic and scientific occupations and organizations, appears in over fifty different scholarly and scientific books and journals. She was named Sociologists for Women in Society (SWS) Feminist Lecturer (for "a prominent feminist scholar who has made a commitment to social change") in 2000 and awarded the Women in Engineering Programs (WEPAN) Betty Vetter Research Award ("for notable achievement in research on women in engineering") in 2002.

DEBORAH G. JOHNSON is the Anne Shirley Carter Olsson Professor of Applied Ethics in the Department of Science, Technology, and Society within the School of Engineering and Applied Science at the University of Virginia. With research interests in the connections between technology and ethics, Johnson has authored or edited four books on computers, ethics, and social values and engineering ethics.

LINDA L. LAYNE is Hale Professor of Humanities and Social Sciences at Rensselaer Polytechnic Institute in Troy, N.Y. She is now co-producing a television series with George Mason University, *Motherhood Lost: Conversations,* which promotes a women's health approach to pregnancy loss. Current research interests include design criteria for feminist technologies and barriers to women's advancement in academe.

CHERYL B. LEGGON is an associate professor at Georgia Institute of Technology in the School of Public Policy. Her research focuses on underrepresented groups in the U.S. science and engineering workforce—especially women of color. Before joining Georgia Tech's faculty in 2002, she was director of women's studies and an associate professor of sociology at Wake Forest University.

SUSAN G. MILLER is an associate professor in the College of Technology, Department of Computer Graphics at Purdue University. She received her master of science from Purdue University and her bachelor of science in industrial design from the Ohio State University with an emphasis in visual communication design.

SUE V. ROSSER, who holds a Ph.D. in zoology from the University of Wisconsin, is the dean of the Ivan Allen College of Liberal Arts at Georgia Tech, where she is also a professor of history, technology, and society and of public policy. She has written nine books and approximately 115 journal articles on theoretical and applied issues of women, science, health, and technology.

BARBARA KATZ ROTHMAN, is a professor of sociology at the City University of New York and is the author of a number of books related to gender and technology, including *In Labor: Women and Power in the Birthplace; The Tentative Pregnancy: How Amniocentesis Changes the Experience of Pregnancy; Recreating Motherhood;* and *The Book of Life: A Social and Ethical Guide to Race, Normality, and the Implications of the Human Genome Project.* Her most recent book is *Weaving a Family: Untangling Race and Adoption.*

JUDY WAJCMAN is a professor of sociology in the Research School of Social Sciences at the Australian National University and was formerly a Centennial Professor at the London School of Economics. Her books include *Feminism Confronts Technology, The Social Shaping of Technology,* and *Managing Like a Man: Women and Men in Corporate Management.* Her latest books are entitled *TechnoFeminism* and *The Politics of Working Life* (co-authored with Paul Edward).

MARA H. WASBURN is an assistant professor of organizational leadership in the College of Technology at Purdue University. Her research and consulting focus on mentoring, particularly as it pertains to women in technology. She has developed the new mentoring model Strategic Collaboration ™, designed to increase women's access to the career-enhancing benefits of mentoring.

Index

WOMEN, GENDER, AND TECHNOLOGY

Women, Gender, and Technology *Edited by Mary Frank Fox, Deborah G. Johnson, and Sue V. Rosser*

The University of Illinois Press
is a founding member of the
Association of American University Presses.

Composed in 10.5/13 Adobe Minion
with Meta display
by Jim Proefrock
at the University of Illinois Press
Manufactured by Thomson-Shore, Inc.

University of Illinois Press
1325 South Oak Street
Champaign, IL 61820-6903
www.press.uillinois.edu